Anonymous

The Metric System of Weights and Measures

Anonymous

The Metric System of Weights and Measures

ISBN/EAN: 9783337338947

Printed in Europe, USA, Canada, Australia, Japan

Cover: Foto ©berggeist007 / pixelio.de

More available books at **www.hansebooks.com**

THE

METRIC SYSTEM

OF

WEIGHTS AND MEASURES.

ISSUED BY

THE HARTFORD STEAM BOILER INSPECTION
AND INSURANCE COMPANY.

HARTFORD, CONN., U. S. A.
1898.

PRESS OF THE CASE, LOCKWOOD & BRAINARD CO.

PREFACE.

THE Metric System of Weights and Measures is used so universally in foreign books and periodicals, that much time is consumed and no little annoyance incurred by the American reader, in translating these units into their English and American equivalents, by the aid of any of the reduction tables that have yet been published. It therefore occurred to the undersigned that a handy pocket volume, for facilitating comparisons of this kind, might be acceptable to engineers and scientific workers generally. This is the *raison d'être* of the present little book. The work of calculating the tables and reading the proof has been done by Mr. A. D. Risteen, Associate Editor of THE LOCOMOTIVE, whose experience in the computing division of the United States Coast and Geodetic Survey should be a sufficient guarantee of the care and accuracy with which his task has been done. Mr. Risteen has also prepared a brief account of the inception and adoption of the Metric System in France, which adds materially to the

(3)

interest and value of the volume. The Metric System is now in common use in France, Germany, Austria-Hungary, Italy, Spain, Portugal, Sweden, Norway, Belgium, The Netherlands, Roumania, Servia, Egypt, Mexico, Costa Rica, Guatemala, Nicaragua, Salvador, Brazil, Colombia, Bolivia, Chile, The Argentine Republic, Haiti, and Santo Domingo. It has also been adopted officially by Venezuela, Uruguay, Turkey, and British India, although in these four countries it can hardly yet be said to be in common use. Japan, too, has announced her intention of adopting a modified form of it, and even China has a *decimal* system of weights and measures, although it is not the French Metric System. England, Russia, and the United States are the only great nations that still refuse to adopt the Metric System. Its advantages are many, and the only really serious objection appears to be, that the change from our present units to the new ones would be more or less confusing and annoying for the first few years. Much of this annoyance and confusion could be prevented by providing school children with cheap sets of metric measures and weights, and requiring each child to measure and weigh a certain number of objects every week. In this way the units and their names would become tolerably familiar to the next generation, and the transition would be far easier. The formidable decimals that the metric system suggests to the average citizen (and which are here very much in evidence) constitute no part of the system itself, and they could be gratefully dismissed, the moment the metric system came into use. They owe their existence to the fact that the foot, pound, and quart are not commensurable with the meter, gramme, and liter; and, when we cease to use our present units, we should also cease to use the long numbers that express their values in metric units.

J. M. ALLEN, *President.*

Hartford, Conn., May 2, 1898.

THE METRIC SYSTEM OF WEIGHTS AND MEASURES.

DIVERSITY OF SYSTEMS IN USE.

From the earliest days of civilization men have measured and weighed things, and hence they have had to have at least the rudiments of a system of weights and measures. At first, however, there was no attempt at precision in the standards used. A "foot" meant something of substantially the same length as the king's foot; and an "inch" was sometimes considered to be the twelfth part of a foot, and at other times it was defined as the space covered by three good plump kernels of barley, placed end to end. Each community had its own units, which in most cases bore no relation whatever to the units employed in other communities. As wealth increased and commerce grew, it became necessary to provide for a greater degree of precision in standards and for a comparison of those used in different localities. Undoubtedly, the innate dishonesty of man also had something to

do with the establishment of more definite
standards, because in the absence of these
there was nothing to prevent a person who
was so inclined from buying by a long
measure and selling by a short one. At
all events, at the present time the units
used in weighing and measuring have
been defined by law in all civilized countries. This has secured, for each country,
a certain degree of uniformity throughout
its own territory; but among the standards of *different* countries there is a diver-
sity almost as great as ever.

Even in the United States there are an
enormous number of units in vogue. For
example, for measuring lengths we have
the inch, ell, nail, link, foot, yard, fathom,
rod, chain, furlong, mile, knot, and doubt-
less many others; and the units in use for
area, capacity, and weight are even more
numerous. Many of these units have different values, too, according to the nature
of the substance to be measured or weighed.
Thus we have several kinds of ounces and
pounds, and at least two kinds of quarts.
There is no regularity whatever in the
number of units of one kind that make one
unit of the next higher kind; and some of
the tables that the schoolboy has to learn
are extremely distressing. We well re-
member what a stickler it was, in our day,

that it should take 30¼ square yards to make one square rod. If we had a simpler method of mensuration, the children in our schools could be spared a great part of this troublesome lumber, and the time saved in this way could be devoted to some far more profitable subject. Taking everything into consideration, it is certainly humiliating to reflect upon the heterogeneous "system" of weights and measures in use at the present day in a country as proud, as civilized, and as advanced (in other respects) as our own United States of America. We are not unique in this respect, it is true, for many other nations have "systems" of measurement that are fully as confusing as our own ; but this is mighty cold comfort.

HISTORY OF THE METER.

To France belongs the honor of making the first systematic attempt to break through the customs of antiquity, and to substitute a new metrology for the old. Until the latter part of the eighteenth century there was the same condition of affairs in that country that prevails to-day in the United States; and in Méchain and Delambre's *Base du Système métrique decimal* (Paris, 1806), we read of "le système incohérent de nos mesures," "l'étonnante

et scandaleuse diversité de nos mesures,"
etc. Several ineffectual attempts to reform the French system of weights and
measures had been made, previous to 1790,*
but it is from that year that the present
"metric system" dates. In May, 1790, M.
de Talleyrand proposed to the National
Assembly of France that a new system of
measures should be devised on strictly
scientific principles, and that the units of
length and weight in this system should
be based on some natural and invariable
standard.   On the 8th of May the Assembly
passed a resolution requesting the king,
Louis XVI, to open a correspondence on the

---

* "Plus d'une fois on avoit présenté des projets de
réferme au gouvernement, qui les avoit fait examiner :
mais, malgré les rapports les plus favorables, malgré
la bonne volonté des ministres, et particulièrement du
contrôleur général des finances Orry, ces projets avoi-
ent toujours été repoussés ou mis en oubli. En 1788, le
vœu d'une mesure uniforme fut consigné dans les
cahiers de quelques balliages ; quelques savans firent
entendre leur voix. Les esprits étoient alors disposés ·
à recevoir avec enthousiasme toutes les réfermes
utiles. Le système incohérent de nos mesures, outre
ses inconvéniens réels, avoit un vice originel qui en fit
hâter l'abolition : la confusion qui y régnoit étoit en
grande partie l'ouvrage de cette féodalité que personne
n'osoit plus défendre, et dont on travailloit à faire dis-
parôitre jusqu'aux moindres vestiges. Ce concours
unique de circonstances valut en accueil favorable à
la proposition faite en 1790 à l'assemblée constituante
par M. de Talleyrand." — Méchain et Delambre, *Base
du Système mètrique decimal.*

subject with the king of England, desiring
him to invite the British Parliament to co-
operate with the National Assembly of
France in fixing the "natural unit" which
was to serve as the basis of the proposed
system of weights and measures. The
work was to be put in the hands of a joint
commission of scientific men, half of whom
were to be appointed by the French Academy of Sciences, and the other half by the
Royal Society of London. In conformity
with the resolution, Louis laid the matter
before the English king; but "owing to the
temper and the public troubles of the
times, his overture met with no response.
Similar applications to other nations were
more successful, and in subsequent proceedings Spain, Italy, the Netherlands,
Switzerland, Denmark, and Sweden, par-
ticipated by sending delegates to an international commission. The system itself
was, however, matured by the labors of a
committee of the Academy of Sciences,
embracing Borda, Lagrange, Laplace,
Mongé, and Condorcet, five of the ablest
mathematicians of Europe."* Lavoisier,
though not a member of this committee,
contributed largely to its proceedings, and
the standards that were afterwards pre-

* Johnson's Cyclopædia, article *Metric System.*

pared were made under his supervision. Lavoisier, for some reason, has not received the recognition that is due him in this matter, his work being usually credited to Borda.

There were two reasons advanced for basing the new unit of length upon some absolute and invariable quantity in nature. It was said that the selection of an arbitrary unit would violate the most fundamental principle of the proposed reform ; for the one distinctive feature of the new system was to be the exclusion of the last vestige of arbitrariness. This argument seems to us to be hardly worth serious consideration, for however satisfactory the unit finally chosen might be, the *selection* of that unit would necessarily be arbitrary, and hence the unit itself would be so, in some sense. The other reason advanced for the selection of a natural unit was much more logical. It was urged that a system of weights and measures based on some permanent and invariable natural quantity could be entirely reconstructed, with any desired degree of precision, even though every standard in existence were utterly destroyed.

M. de Talleyrand proposed to adopt, as the fundamental unit of length, the length of a pendulum vibrating seconds in lati-

tude 45°, "ou toute autre latitude qui pour-
roit être préférée" (or such other latitude
as might be preferred). The committee of
the Academy of Sciences considered the
advisability of adopting the seconds pendulum as the unit of length, but finally
rejected it, principally because it involved
the conception of *time*. It seemed to them
preferable to base the proposed unit on
*the length of some object actually existing in
nature;* and after much deliberation it
was decided to adopt some one of the dimensions of the earth itself. Of the different dimensions proposed, the two that met
with the most favor were the equatorial
circumference, and the meridian quadrant;
and of these two the meridian quadrant
was finally selected, because it could be
measured with greater áccuracy. (The
measurement of a *meridian* involves observations of *latitude*, while the measurement
of the *equator* involves observations of
longitude; and before the invention of the
electric telegraph, longitude measures
were subject to large errors.) In order to
guard against possible differences in the
lengths of the various meridian quadrants,
it was decided to recommend that the new
system of measures be based on the par-
ticular meridian that passes through Paris.
This length (about 6,000 miles) would be

entirely out of the question as a practical
unit for business purposes.   However, a
rough calculation showed that a standard
having a length equal to one ten-millionth
part of this quadrant would be convenient,
and the METER (as the new unit was named)

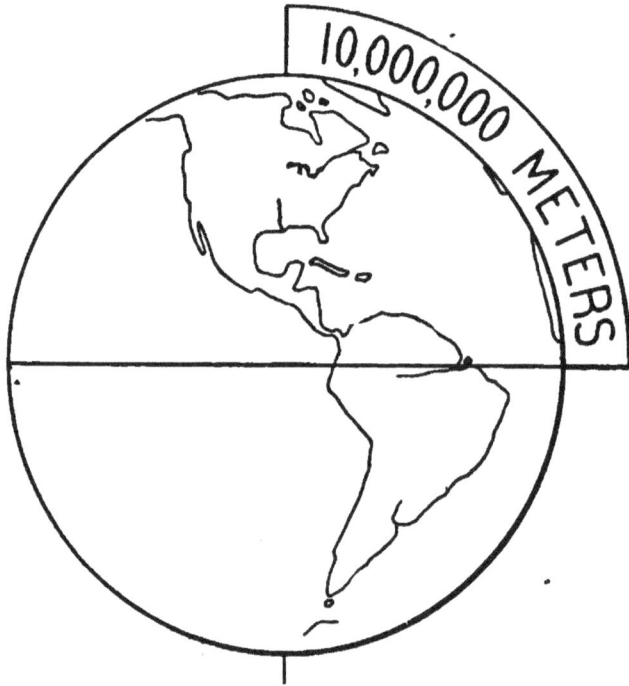

FIG. 1.— THE DEFINITION OF THE METER.

was therefore defined to be *the ten-millionth
part of the distance from the equator to the
north pole, measured along the sea-level, on
the meridian passing through Paris.*   (Fig. 1
illustrates this definition of the meter very
graphically.)

The report of the committee, embodying the points explained above, and dated March 19, 1791, was approved by the Na-

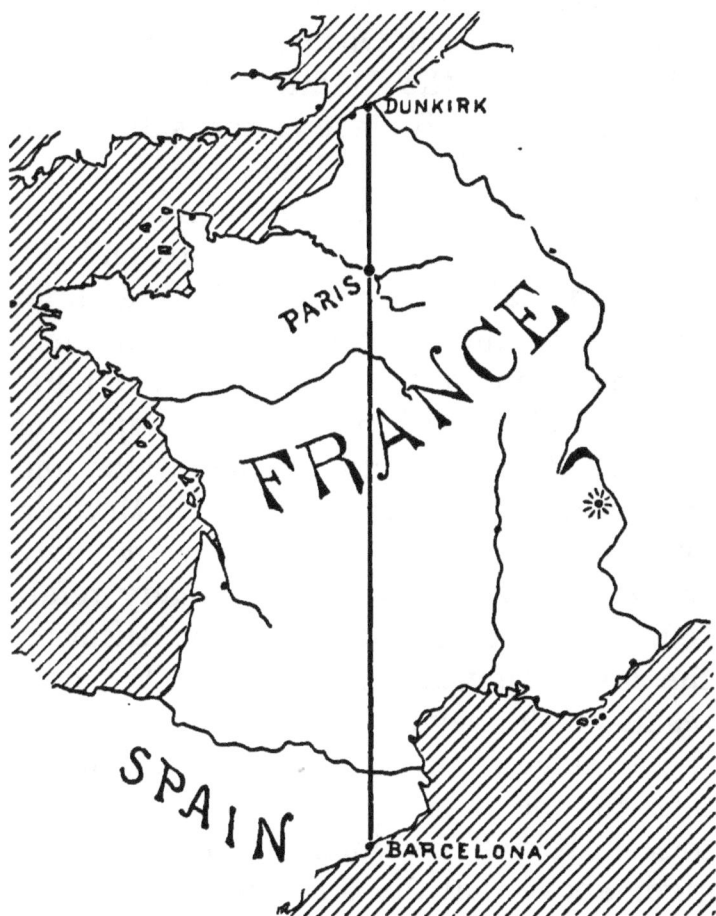

FIG. 2.— SHOWING THE LINE THAT WAS MEASURED.

tional Assembly, and the work of measuring the meridian passing through Paris was begun. This operation involved immense labor, and seven years were required to complete it. The meridian through Paris is shown by the heavy vertical line in Fig. 2. It strikes the North Sea at Dunkirk, near the Strait of Dover, in latitude 51° 02' 8.85", and the Mediterranean Sea at Montjouy (a small place in the suburbs of Barcelona, Spain), in latitude 41° 21' 44.96". The northern half of this arc was measured under the direction of Delambre, and the southern half under Méchain. A brief account of the method of measurement employed, and of the subsequent computations, will be found in the first chapter of Clarke's *Geodesy;* and the operations and calculations are described in full in Méchain and Delambre's *Base du Système métrique décimal,* to which we have already referred. The unit of length used in measuring the meridian was the *toise,* and the final result of the seven years' work was, that the distance from the Barcelona end of the line to the Dunkirk end was 551,584.7 toises. It will be found from the data given above that the difference in latitude between the two ends was 9° 40' 23.89". Now the latitude of the equator is 0°, and the latitude of the pole is 90° ; and therefore if the earth were a true sphere the distance from the equator to the pole would be given by the simple proportion

9° 40' 23.89'' : 90° :: 551,584.7 toises : distance
    required.

The first term of this proportion, when
reduced to degrees and decimals of a
degree, is 9.6733° ; and if we substitute this
value of it, and then solve the proportion
by the ordinary rule-of-three, we find

Distance from equator to pole = 5,131,922 toises.

In the actual calculation a small allowance
had to be made for the fact that the earth
is not a perfect sphere (the polar diameter
being about 26 miles shorter than the equatorial diameter). It was found that to take
this into account it would be necessary
to subtract 1,181 toises from the distance
as calculated above. Hence, it was con-
cluded that the true distance from the
equator to the pole, along the meridian of
Paris, is 5,130,741 toises. One ten-millionth
part of this is 0.5130741 of a toise, and this,
therefore, was to be the length of the new
unit, or *meter*.

The next task was to prepare a bar of
platinum which should have this length.
When this had been done, the length of
the bar was verified by the international
commission referred to above, and the
commission then proceeded in a body to
the Palace of the Archives, in Paris, and
there they formally deposited the bar of

platinum which was to be ever afterward the standard meter of the world.

UNITS DERIVED FROM THE METER.
LONG MEASURE.

The meter is theoretically equal to the ten-millionth part of the meridian-quadrant that passes through Paris ; but as all human measurements are liable to error, we cannot suppose that Méchain and Delambre's determination of this length is absolutely exact, and therefore for practical purposes the meter is defined to be *the length of the platinum bar that was deposited in the Palace of the Archives by the International Commission.* A comparison of this bar with the units in use in this country shows that the meter is about 3⅜ inches longer than a yard-stick. A more exact comparison, made by Professor Rogers, indicates that the precise length of the standard meter is 39.37027 inches ; and a similar determination by General Comstock gave 39.36985 inches. The average of these two determinations is 39.37006 inches, which is the value adopted in this essay as the true length of the meter.

To make the new system available for general purposes, it was decided to divide the meter into ten equal parts, which were called *deci*meters. Each of these parts was further subdivided into ten parts, called *centi*meters, and each centimeter was also divided into ten parts, called *milli*meters. (The italicised prefixes stand for "tenth," "hundredth," and "thousandth," respectively, just as the words "dime," "cent," and "mill" stand for the tenth, hundredth, and thousandth part of a dollar. We have a decimal system of coinage, and we realize its simplicity and general utility. Why not apply the same idea to our weights and measures?) Fig. 3 is a DECIMETER. The decimeter is shown subdivided into ten CENTIMETERS, and each centimeter is further subdivided into ten MILLIMETERS. Ten of these decimeters, placed end to end, would make *one meter*; and 100,000,000 of them, if laid end to end along the curved surface of the earth, would just reach from the equator to the north pole.

For measuring big distances it was proposed to establish a special unit equal to 1,000 meters. To this unit the name KILOMETER was given ("kilo" being derived from a Greek word meaning "a thousand"). The meter being equal to 39.37006 inches, it follows that the KILO-METER is equal to 39,370.06 inches — that is, to 3,280.838 feet, or 0.6213709 of a mile.

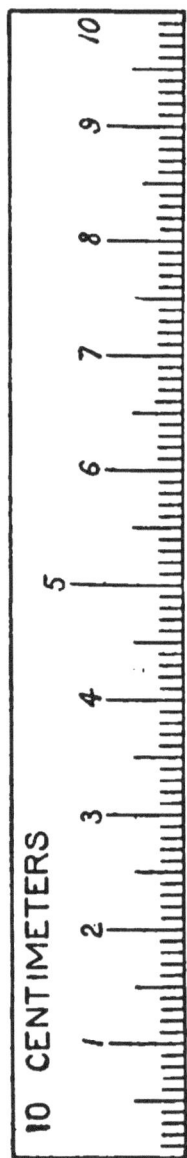

10 CENTIMETERS

1 DECIMETER.

This scale is 1 decimeter in length, and is subdivided into 10 centimeters, and again into 100 millimeters. Since the decimeter is the tenth part of a meter, 100,000,000 of these scales, placed end to end along the meridian passing through Paris, would just reach from the equator to the north pole. A cubical box, whose inside dimensions were all equal to the length of this scale, would contain 1 liter. A kilogram of water would just fill such a box, at 39° Fahr. At this same temperature, a cube of water, 1 centimeter long, high and wide, would weigh 1 gramme.

The table for long measure, in the Metric System, is very simple. It is as follows :

10 millimeters (mm.)* = 1 centimeter.
10 centimeters (cm.) = 1 decimeter.
10 decimeters (dm.) = 1 meter.
1,000 meters (m.) = 1 kilometer (km.).

SQUARE MEASURE.

The tables for square measure and cubic measure are easily derived from this table for long measure. The area of a rectangle or square, for example, is equal to the product of the length and breadth. A square 1 decimeter (or 10 centimeters) each way will therefore have an area of 100 square centimeters : hence 1 square decimeter = 100 square centimeters. The rest of the table for square measure is constructed in the same way, and hence we have the following table :

100 square millimeters (sq. mm.) = 1 square centimeter.

100 square centimeters (sq. cm.) = 1 square decimeter.

100 square decimeters (sq. dm.) = 1 square meter.

1,000,000 square meters (sq. m.) = 1 square kilometer (sq. km.).

---

* The letters in parentheses indicate the customary abbreviation for the units opposite which they stand.

CUBIC MEASURE.

The table for cubic measure is constructed in the same way, by remembering that the volume of a cube is equal to the product of the length, breadth, and height. Thus a cubic meter, being 10 decimeters each way, will have the same volume as 1,000 cubic decimeters. Hence, we have the following table :

1,000 cubic millimeters (c. mm.) = 1 cubic centimeter.

1,000 cubic centimeters (c. c.) = 1 cubic decimeter.

1,000 cubic decimeters (c. dm.) = 1 cubic meter.

1,000,000,000 cubic meters (c. m.) = 1 cubic kilometer (c. km.).

The cubic meter was formerly called a *stere*, and the units derived from it were called *decisteres, centisteres*, etc. These names are now practically out of use, those given in the foregoing table being generally preferred.

MEASURES OF MASS.

The unit of *mass*, in the metric system, is the GRAMME. A gramme is the quantity of matter in a cubic centimeter of pure water, at the temperature of its maximum density (about 39° Fah.). It might seem more natural to take, as the unit of mass, the quantity of matter in a cubic centimeter of water at the *freezing point.* There were two reasons why this was not done. In the first place, it is difficult, in experimenting with water at the exact freezing point, to make sure that the water does not contain minute needles of ice — it certainly will contain them if any part of the water under examination is exposed, even momentarily, to a temperature that is the least fraction of a degree *below* the freezing point — and the presence of such particles of ice would modify the density of the water to such an extent that it would be impossible to determine the value of the gramme with the requisite degree of precision. The second reason for preferring the temperature of maximum density in defining the gramme is, that at this point the coefficient of expansion of water is zero, so that a small change in temperature produces no sensible variation in the density of the water.

Other units of mass were derived from the gramme, in precisely the same way that other units of length were derived from the meter. Thus, the tenth part of a gramme was called a *deci*gram, the hundredth part of a gramme was called a *centi*gram, the thousandth part of a gramme was called a *milli*gram, and a weight equal to a thousand grammes was called a *kilo-*

gram. The kilogram is therefore theoretically equal to the *mass of a cubic decimeter of pure water, at its temperature of maximum density.* The weight of this quantity of water was determined experimentally by the Commission that prepared the standard platinum meter, and a standard mass of platinum, having this same weight as nearly as the unavoidable errors of experiment would allow, was deposited in the Archives of France, by the Commission, on June 22, 1799. For practical and legal purposes, therefore, *the kilogram is defined as the weight, in* vacuo, *of this standard mass of platinum.* Miller's comparison of this weight with the equivalents used in our own country indicates that the KILOGRAM is equal to 2.20462125 AVOIRDUPOIS POUNDS. The avoirdupois pound being equal to 7,000 troy grains, it follows that the GRAMME is equal to 15.43234875 TROY GRAINS. The table for weights in the Metric System is as follows:

10 milligrams (mgm.)   = 1 centigram.
10 centigrams (cgm.)   = 1 decigram.
10 decigrams (dgm.)   = 1 gramme.
1,000 grammes (gm.)   = 1 kilogram.
1,000 kilograms (kgm.) = 1 millier.*

It will be noticed that we have preserved the French spelling in the case of the

---

*Also called "tonne, "or "tonneau."

gramme, although the final *me* has been dropped from the other words. This is in conformity with the recommendation of the Committee appointed by the American Association for the Advancement of Science, for considering the revision of the spelling of words used in chemistry. The *me* is dropped from the other words for the sake of simplicity, but it is preserved in the case of the *gramme* to avoid confusion with the word *grain*. The reader will see the importance of this if he will hastily write down the words *gram* and *grain*.

LIQUID MEASURE.

For measuring liquids the Commission provided a unit called a LITER, which is not greatly different from a United States liquid quart. The liter is defined as *the volume occupied by a kilogram of water at the temperature of maximum density*. For all practical purposes this is equivalent to saying that the LITER is equal to ONE CUBIC DECIMETER. A comparison with the corresponding United States measures shows that the liter is equal to 1.0563 liquid quarts, or 0.9081 dry quarts. The table for liquid measure, in the Metric System, is derived from the liter in precisely the same way as the table for long measure is derived from the meter. Thus we have

10 milliliters (ml.) = 1 centiliter
10 centiliters (cl.) = 1 deciliter
10 deciliters (dl.) = 1 liter
1,000 liters (l.)   = 1 kiloliter (kl.)

There is no distinction, in the Metric System, between "wet" and "dry" measure.

LAND MEASURE.

For convenience in measuring *land*, the unit called the ARE was proposed. The ARE is equal to the area of a square whose edges are each 10 meters long; and hence it is equivalent to 0.024711 of an acre. The table for land measure is as follows:

10 milliares (ma.) = 1 centare
10 centares (ca.) = 1 deciare
10 deciares (da.) = 1 are
10 ares (a.)   = 1 decare
10 decares (dc.)   = 1 hectare (hc.)

The HECTARE, equal to 2.4711 acres, is the unit commonly employed in measuring tracts of land of any considerable size.

MISCELLANEOUS UNITS.

In addition to the units of measurement that have already been mentioned, certain compound units have been tabulated in the pages that follow. It is believed that these will be almost self-explanatory. The English and American unit for the measurement of WORK is the FOOT-POUND, which is the work that must be done to raise a pound through a height of one foot. The corresponding metric unit is the KILOGRAM-METER, which is the work that must be done in order to raise a weight of one kilogram through a height of one meter. Since a kilogram is equal to 2.20462 pounds, and a meter is equal to 3.28084 feet, it follows that one KILOGRAM-METER is equal to $2.20462 \times 3.28084 = 7.23300$ FOOT-POUNDS. The other compound units are derived in the same simple way. The only units that still call for special mention appear to be those used for the measurement of *heat* and *power*. The English unit of heat, is the quantity of heat required to raise the temperature of one pound of water one degree on the Fahrenheit scale. The corresponding metric unit is the CALORIE, which is the heat required to raise the temperature of a kilogram of water one degree on the Centigrade scale. A kilogram being equal to 2.20462 pounds, and a Centigrade degree being equal to 1.8 Fahrenheit degrees, it follows that the CALORIE is equal to $1.8 \times 2.20462 = 3.96832$ English heat units. The metric unit for meas-

uring the power of a steam engine or
other prime mover is the FORCE DE CHEVAL,
which is defined as equal to the performing
of 4,500 kilogram-meters per minute.   A
kilogram-meter being equal to 7.233 foot-
pounds, it follows that the FORCE DE CHEVAL
is equal to $4,500 \times 7.233 = 32,549$ foot-pounds
per minute.   The HORSE-POWER being
equal to the performance of 33,000 foot-
pounds of work per minute, it follows that
the FORCE DE CHEVAL is equal to $32,549 \div$
$33,000 = 0.98633$ of a HORSE-POWER.

THERMOMETER SCALES.

The Centigrade thermometer scale,
though not properly a part of the Metric
System, may be mentioned in this place.
On the Fahrenheit scale the freezing point
of water is defined as 32°, and the size of
the degrees is fixed by defining the difference in temperature between the freezing
and boiling points to be 180°. This is a
clumsy sort of scale, and little can be said
in favor of it except that it is in use in
England and the United States.   The
Centigrade scale was determined by defin-
.ing the freezing point of water to be 0°,
and the boiling point to be 100°. The following rules result from these definitions:
To convert a given reading on the Fahrenheit scale to its equivalent on the Centi-

grade scale, subtract 32° and divide the remainder by 1.8. To convert a given reading on the Centigrade scale to its equivalent on the Fahrenheit scale, multiply it by 1.8 and to the product add 32°.

USE OF THE TABLES.

The tables appended hereto are intended to facilitate the comparison of the metric system with the weights and measures used in this country. They are arranged in order, beginning with simple measures of distance, under which head centimeters are compared with inches, meters with feet, and kilometers with miles. The next set of tables refers to measures of *area*, and square inches, square feet, square yards, square miles, and acres, are compared with the corresponding metric measures. *Cubic* measure is next considered, and cubic inches, cubic feet, cubic yards, fluid ounces, dry quarts, liquid quarts, gallons, etc., are tabulated. Measures of *weight* come next, and in this set of tables we find grains, ounces, pounds, tons, etc. The last set of tables contains miscellaneous complex units, such as foot-pounds, pounds-per-square-inch, etc., which are compared with the corresponding metric units.

It is believed that these tables require

but little further explanation. If we open
at any part, we find a table for reducing a given number of units in the metric system to the corresponding units employed in this country; and immediately following we find a similar table
for performing the converse operation —
namely, for reducing a given number of
United States units to the corresponding
metric equivalent. For example, if we
wish to know how many square inches
there are in 38 square centimeters, we
turn to page 60, where we see at once
that 38 square centimeters are equivalent
to 5.89001 square inches. If we wish to
know how many square centimeters there
are in 26 square inches, we look in the
immediately following table, and there
we find that 26 square inches are equivalent to 167.742 square centimeters. It will
be noticed that the tables run only to 100
units of any given kind. If we wish to
know the value of a larger number of
units than this, we may proceed as follows: If the proposed number is com-
posed entirely of ciphers, except for the
first one or two figures, we can deduce
the desired equivalent directly from the
tables, by merely shifting the decimal
point one place to the right for every
cipher that follows the significant figures

with which the proposed number begins.
Thus, suppose we wished to reduce 980
cubic meters to its equivalent in cubic
yards. The table that we have to use
is given on pages 88 and 89. We find,
here, that 98 cubic meters=128.1790 cubic
yards ; and hence 980 cubic meters=1281.-
790 cubic yards. If we were required to
reduce 4600 cubic meters to cubic yards,
we should find that 46 cubic meters=60.-
1657 cubic yards, and hence 4600 cubic
meters=6,016.57·cubic yards. Similarly,
54,000 cubic meters=70,629.2 cubic yards,
and 770,000 cubic meters=1,007,120. cubic
yards. If the proposed number contains
more than two significant figures, we may
split it up into two or more other numbers,
each of which consists of not more than
two such figures, and then we can deduce
the values of each of these partial numbers,
and by adding the equivalent numbers so
found, we can deduce the equivalent of the
number originally proposed. Thus, suppose we wished to know how many inches
there are in 4,189 centimeters. (The table
that we have to use is given on pages 44
and 45.) We see that 4,189=4,100+89.
Hence the operation is as follows :

$$4100 \text{ cm.} = 1614.172 \text{ in.}$$
$$89 \text{ cm.} = 35.03935 \text{ in.}$$
$$\therefore 4189 \text{ cm.} = 1649.21135 \text{ cm.}$$

As the table does not give us the equivalent of 4100 centimeters to more than
three decimal places, the value of 4189
centimeters, as obtained above, cannot be
regarded as correct to more than three
places of decimals. Hence we reject the
last two decimals in the result already
found, and we have

4189 centimeters=1649.211 inches.

Of course we may divide up the proposed
number in any way we please, for the pur-
pose of resolving it into other numbers of
two significant figures each. We wrote
4189=4100+89, merely because that is the
easiest decomposition of this sort that can
be found. But we might have decomposed
it in any other manner. For example, we
might split it up this way :

4189=3800+290+43+56.

Then the work would be :

3800 cm. =1496.062	in.
290 cm.= 114.1732	in.
43 cm.= 16.92913	in.
56 cm.= 22.04723	in.

∴ 4189 cm. =1649.21156 in.

Here, as before, we must regard the last
two figures as inaccurate, because the
table does not give us the value of 3800

centimeters to more than three places of decimals. The result obtained by splitting up a proposed number in different ways will sometimes differ among themselves by a unit in the last decimal place. This arises from the fact that only a limited number of decimal places have been preserved in the tables.

This source of apparent error can never be avoided in tabulating incommensurable quantities, for no matter how many decimal places are preserved, the last one is necessarily too great or too small, and the sum of two or more numbers taken from the table is liable to be slightly in error in the last place. If the table is accurate the error thus introduced can never exceed *one* unit in the last place when *two* tabular numbers are added ; and, in general, it cannot exceed $\frac{1}{2}n$ units in the last place when n numbers are added. In the present tables, it is believed that the decimal places have been carried out far enough to ensure that errors of this sort, in practice, will have no important effect.

The examples that have now been given will suffice to illustrate the use of all the tables except those on pages 188 to 189, where the Fahrenheit and Centigrade thermometers are compared. The engravings

on pages 190 to 192 will give, at a glance, a
good idea of how these two scales compare
at different points; and the tables enable
us to calculate, to the nearest hundredth
of a degree, the reading on either one of
these scales, which corresponds to a given
reading on the other one. The rule for performing this calculation *without* the tables,
has been given already, on page 28. It
only remains to show how the calculation
is facilitated by the use of the tables. The
table on page 188 gives a direct comparison
for every five degrees on the Centigrade
scale up to 230°, and then for every ten
degrees up to 470°. Suppose that a Fahrenheit thermometer and a Centigrade
thermometer are exposed, side by side, to
a given temperature. If the Centigrade
instrument reads 170°, we see from the
table that the Fahrenheit instrument will
read 338°; if the Centigrade thermometer
reads 0°, the Fahrenheit one will read 32°;
if the Fahrenheit one reads 212°, the Centigrade one will read 100°; and if the Fahrenheit one reads 788°, the Centigrade one
will read 420°. A given temperature on
either scale may thus be translated directly
into its equivalent on the other one, provided the given temperature can be found
in the table on page 188. If it cannot be
found there, we have to make use of the

auxiliary table on page 189, as illustrated in the examples which follow. Suppose the Centigrade instrument reads 42°, and we wish to know the corresponding reading on the Fahrenheit one. We do not find 42° in the Centigrade column, but we *do* find 40°, and its equivalent in the Fahrenheit system is 104°. Turning now to the auxiliary table on the next page, we see that a 2° interval on a Centigrade thermometer is equivalent to an interval of 3.6° on a Fahrenheit instrument. Hence we obtain the equivalent of 42° C., as follows:

$$40° \text{ C.} = 104.° \text{ Fahr. (p. 188).}$$
$$\text{Correction for } 2° \text{ C.} = \underline{\quad 3.6° \quad} \text{ `` (p. 189).}$$
$$\text{Hence, } 42° \text{ C.} = 107.6° \text{ Fahr.}$$

We might have based our calculation on any other reading in the main table, which is within 20° of the given temperature of 42° (the 20° being the greatest interval tabulated on page 189). Thus, we could base it upon the reading 55° in the first column. We note that 42° = 55° — 13°, and the calculation is as follows:

$$55° \text{ C.} = 131.° \text{ Fahr. (p. 188).}$$
$$\text{Correction for } 13° \text{ C.} = \underline{\quad 23.4° \quad} \text{ `` (p. 189).}$$
$$\text{Hence, } 42° \text{ C.} = 107.6° \text{ Fahr.}$$

which is the same result as we obtained before. By way of completing the explanation of these tables, let us suppose that it is desired to reduce the Fahrenheit temperature 703.89° to its equivalent on the Centigrade scale. The work is as follows:

$$703.89° = 698° + 5° + .8° + .09°$$

Then

698.° Fahr.=370.°   C. (p. 188).
Correction for 5.°   "   =   2.78° C. (p. 189).
    "   " .8° "   =   .444°C.   "
    "   " .09° "   =   .050°C.   "

Hence, 703.89° Fahr. = 373.274°C.

But since the correction for 5° Fahr. is only given to two places of decimals, the *third* decimal in the result cannot have any pretension to accuracy; so that, preserving the nearest unit in the *second* decimal place, our conclusion is, that when a Fahrenheit thermometer reads 703.89°, a Centigrade thermometer, exposed to the same conditions, will read 373.27°.

DATA

Used in Computing the Tables.

Metric Unit.	U. S. Equivalent.	Logarithm.
1 Centimeter	0.393 700 6 inch.	9.595 1661
1 Meter	3.280 838 feet.	0.515 9849
1 Meter	1.093 613 yards.	0.038 8636
1 Kilometer	0.621 370 9 mile.	9.793 3510
1 Square centimeter	0.155 000 2 square inch.	9.190 3322
1 Square meter	10.763 90 square feet.	1.031 9698
1 Square meter	1.195 989 square yards.	0.077 7272
1 Square kilometer	0 386 101 8 square mile.	9.586 7020
1 Hectare	2.471 052 acres.	0.392 8820
1 Cubic centimeter	0.061 023 66 cubic inch.	8.785 4983
1 Cubic meter	35.314 63 cubic feet.	1.547 9547
1 Cubic meter	1.307 949 cubic yards.	0.116 5908
1 Cubic kilometer	0.239 912 4 cubic miles.	9.380 0530
1 Cubic centimeter	0.033 802 2 U. S. fluid ounce.	8.528 9445
1 Liter	1.056 304 U. S. liquid quarts.	0.023 7888
1 Liter	0.264 076 U. S. gallon.	9.421 7288
1 Liter	0.880 860 British liquid quart.	9.944 9069
1 Liter	0.220 215 British gallon.	9.342 8469

Metric Unit.	U. S. Equivalent.		Logarithm.
1 Liter	0.908 082	U. S. dry quart.	9.958 1250
1 Stere (cubic meter)	28.377 6	U. S. bushels.	1.452 9750
1 Gramme	15.432 348 7	grains.	1.188 4320
1 Gramme	0.035 273 94	avoirdupois ounce.	8.547 4540
1 Kilogram	2.204 621 2	avoirdupois pounds.	0.343 3340
1 Millier (or tonneau)	1.102 310 7	short tons (2000 lbs.)	0.042 3040
1 Gramme	0.032 150 7	Troy (or apoth.) ounce.	8.507 1908
1 Kilogram	2.679 228	Troy (or apoth.) pound.	0.428 0096
1 Kilogram per sq. met'r	0.204 816 1	pound per sq. foot.	9.311 3642
1 Kilogram per sq. cm.	14.223 35	pounds per sq. inch.	1.153 0018
1 Kilogram-meter	7.233 007	foot-pounds.	0.859 3189
1 Calorie	3.968 318	British heat units.	0.598 6065
1 Calorie	3091.36	foot-pounds.	3.490 15
1 Force de cheval	0.986 319 3	horse-power.	9.994 0175
1 Gramme in a cu. cm.	0.578 037 1	ounces in a cu. inch.	9.761 9557
1 Kilogram in a cu. m.	0.062 427 99	pound in a cubic foot.	8.795 3793
1 Millier in a cu. meter	0.842 778 0	short ton in a cu. yard.	9.925 7132
1 Milligram in a liter	0.058 439	grain in a U. S. gallon.	8.766 7032

U. S. Unit.	Metric Equivalent.	Logarithm.
1 Inch	2.540 001 centimeters.	0.404 8339
1 Foot	0.304 800 1 meter.	9.484 0151
1 Yard	0.914 400 meter.	9.961 1364
1 Mile	1.609 344 kilometer.	0.206 6490
1 Square inch	6.451 606 square centimeters.	0.809 6678
1 Square foot	0.092 903 1 square meter.	8.968 0302
1 Square yard	0.836 128 1 square meter.	9.922 2728
1 Square mile	2.589 989 square kilometers.	0.413 2980
1 Acre	0.404 685 8 hectare.	9.607 1180
1 Cubic inch	16.387 083 cubic centimeters.	1.214 5017
1 Cubic foot	0.028 316 9 cubic meter.	8.452 0453
1 Cubic yard	0.764 555 8 cubic meter.	9.883 4092
1 Cubic mile	4.168 186 cubic kilometers.	0.619 9470
1 U. S. fluid ounce	29.583 9 cubic centimeters.	1.471 0555
1 U. S. liquid quart	0.946 698 liter.	9.976 2112
1 U. S. gallon	3.786 79 liters.	0.578 2712
1 British liquid quart	1.135 254 liters.	0.055 0931
1 British gallon	4.541 02 liters.	0.657 1531

U. S. Unit.	Metric Equivalent.		Logarithm.
1 U. S. dry quart	1.101 222	liters.	0.041 8750
1 U. S. bushel	0.035 239 11	stere (cubic meter.)	8.547 0250
1 Grain	0.064 798 95	gramme.	8.811 5680
1 Avoirdupois ounce	28.349 54	grammes.	1.452 5460
1 Avoirdupois pound	0.453 592 6	kilogram.	9.656 6660
1 Short ton (2000 lbs.)	0.907 185 2	millier (or tonneau.)	9.957 6960
1 Troy (or apoth.) ounce	31.103 50	grammes.	1.492 8092
1 Troy (or apoth.) pound	0.373 241 9	kilogram.	9.571 9904
1 Pound per square foot	4.882 427	kilograms per sq. meter.	0.688 6358
1 Pound per square inch	0.070 306 94	kilogram per sq. cm.	8.846 9982
1 Foot-pound	0.138 255 1	kilogram-meter.	9.140 6811
1 British heat unit	0.251 995 9	calorie.	9.401 3935
1 Foot-pound	0.000 323 48	calorie.	6.509 85
1 Horse-power	1.013 871	forces de chevaux.	0.005 9825
1 Ounce in a cubic inch	1.729 993	grammes in a cubic cm.	0.238 0443
1 Pound in a cubic foot	16.018 46	kilograms in a cu. meter.	1.204 6207
1 Short ton in a cu. yard	1.186 552	milliers in a cu. meter.	0.074 2868
1 Grain in a U.S. gallon	17.111 84	milligrams in a liter.	1.233 2968

TABLES

CENTIMETERS TO INCHES.

Centimeters.	Inches.	Centimeters.	Inches.
1	0.39370	26	10.23622
2	0.78740	27	10.62992
3	1.18110	28	11.02362
4	1.57480	29	11.41732
5	1.96850	30	11.81102
6	2.36220	31	12.20472
7	2.75590	32	12.59842
8	3.14960	33	12.99212
9	3.54331	34	13.38582
10	3.93701	35	13.77952
11	4.33071	36	14.17322
12	4.72441	37	14.56692
13	5.11811	38	14.96062
14	5.51181	39	15.35432
15	5.90551	40	15.74802
16	6.29921	41	16.14172
17	6.69291	42	16.53543
18	7.08661	43	16.92913
19	7.48031	44	17.32283
20	7.87401	45	17.71653
21	8.26771	46	18.11023
22	8.66141	47	18.50393
23	9.05511	48	18.89763
24	9.44881	49	19.29133
25	9.84252	50	19.68503

CENTIMETERS TO INCHES.

Centimeters.	Inches.	Centimeters.	Inches.
51	20.07873	76	29.92125
52	20.47243	77	30.31495
53	20.86613	78	30.70865
54	21.25983	79	31.10235
55	21.65353	80	31.49605
56	22.04723	81	31.88975
57	22.44093	82	32.28345
58	22.83463	83	32.67715
59	23.22834	84	33.07085
60	23.62204	85	33.46455
61	24.01574	86	33.85825
62	24.40944	87	34.25195
63	24.80314	88	34.64565
64	25.19684	89	35.03935
65	25.59054	90	35.43305
66	25.98424	91	35.82675
67	26.37794	92	36.22046
68	26.77164	93	36.61416
69	27.16534	94	37.00786
70	27.55904	95	37.40156
71	27.95274	96	37.79526
72	28.34644	97	38.18896
73	28.74014	98	38.58266
74	29.13384	99	38.97636
75	29.52754	100	39.37006

INCHES TO CENTIMETERS.

Inches.	Centimeters.	Inches.	Centimeters.
1	2.5400	26	66.0400
2	5.0800	27	68.5800
3	7.6200	28	71.1200
4	10.1600	29	73.6600
5	12.7000	30	76.2000
6	15.2400	31	78.7400
7	17.7800	32	81.2800
8	20.3200	33	83.8200
9	22.8600	34	86.3600
10	25.4000	35	88.9000
11	27.9400	36	91.4400
12	30.4800	37	93.9800
13	33.0200	38	96.5200
14	35.5600	39	99.0600
15	38.1000	40	101.6000
16	40.6400	41	104.1400
17	43.1800	42	106.6800
18	45.7200	43	109.2200
19	48.2600	44	111.7600
20	50.8000	45	114.3000
21	53.3400	46	116.8400
22	55.8800	47	119.3800
23	58.4200	48	121.9200
24	60.9600	49	124.4600
25	63.5000	50	127.0000

INCHES TO CENTIMETERS.

Inches.	Centimeters.	Inches.	Centimeters.
51	129.5401	76	193.0401
52	132.0801	77	195.5801
53	134.6201	78	198.1201
54	137.1601	79	200.6601
55	139.7001	80	203.2001
56	142.2401	81	205.7401
57	144.7801	82	208.2801
58	147.3201	83	210.8201
59	149.8601	84	213.3601
60	152.4001	85	215.9001
61	154.9401	86	218.4401
62	157.4801	87	220.9801
63	160.0201	88	223.5201
64	162.5601	89	226.0601
65	165.1001	90	228.6001
66	167.6401	91	231.1401
67	170.1801	92	233.6801
68	172.7201	93	236.2201
69	175 2601	94	238.7601
70	177.8001	95	241.3001
71	180.3401	96	243.8401
72	182.8801	97	246.3801
73	185.4201	98	248.9201
74	187.9601	99	251.4601
75	190.5001	100	254.0001

METERS TO FEET.

Meters.	Feet.	Meters.	Feet.
1	3.281	26	85.302
2	6.562	27	88.583
3	9.843	28	91.863
4	13.123	29	95.144
5	16.404	30	98.425
6	19.685	31	101.706
7	22.966	32	104.987
8	26.247	33	108.268
9	29.528	34	111.548
10	32.808	35	114.829
11	36.089	36	118.110
12	39.370	37	121.391
13	42.651	38	124.672
14	45.932	39	127.953
15	49.213	40	131.234
16	52.493	41	134.514
17	55.774	42	137.795
18	59.055	43	141.076
19	62.336	44	144.357
20	65.617	45	147.638
21	68.898	46	150.919
22	72.178	47	154.199
23	75.459	48	157.480
24	78.740	49	160.761
25	82.021	50	164.042

METERS TO FEET.

Meters.	Feet.	Meters.	Feet.
51	167.323	76	249.344
52	170.604	77	252.625
53	173.884	78	255.905
54	177.165	79	259.186
55	180.446	80	262.467
56	183.727	81	265.748
57	187.008	82	269.029
58	190.289	83	272.310
59	193.569	84	275.590
60	196.850	85	278.871
61	200.131	86	282.152
62	203.412	87	285.433
63	206.693	88	288.714
64	209.974	89	291.995
65	213.254	90	295.275
66	216.535	91	298.556
67	219.816	92	301.837
68	223.097	93	305.118
69	226.378	94	308.399
70	229.659	95	311.680
71	232.939	96	314.960
72	236.220	97	318.241
73	239.501	98	321.522
74	242.782	99	324.803
75	246.063	100	328.084

FEET TO METERS.

Feet.	Meters.	Feet.	Meters.
1	0.3048	26	7.9248
2	0.6096	27	8.2296
3	0.9144	28	8.5344
4	1.2192	29	8.8392
5	1.5240	30	9.1440
6	1.8288	31	9.4488
7	2.1336	32	9.7536
8	2.4384	33	10.0584
9	2.7432	34	10.3632
10	3.0480	35	10.6680
11	3.3528	36	10.9728
12	3.6576	37	11.2776
13	3.9624	38	11.5824
14	4.2672	39	11.8872
15	4.5720	40	12.1920
16	4.8768	41	12.4968
17	5.1816	42	12.8016
18	5.4864	43	13.1064
19	5.7912	44	13.4112
20	6.0960	45	13.7160
21	6.4008	46	14.0208
22	6.7056	47	14.3256
23	7.0104	48	14.6304
24	7.3152	49	14.9352
25	7.6200	50	15.2400

FEET TO METERS.

Feet.	Meters.	Feet.	Meters.
51	15.5448	76	23.1648
52	15.8496	77	23.4696
53	16.1544	78	23.7744
54	16.4592	79	24.0792
55	16.7640	80	24.3840
56	17.0688	81	24.6888
57	17.3736	82	24.9936
58	17.6784	83	25.2984
59	17.9832	84	25.6032
60	18.2880	85	25.9080
61	18.5928	86	26.2128
62	18.8976	87	26.5176
63	19.2024	88	26.8224
64	19.5072	89	27.1272
65	19.8120	90	27.4320
66	20.1168	91	27.7368
67	20.4216	92	28.0416
68	20.7264	93	28.3464
69	21.0312	94	28.6512
70	21.3360	95	28.9560
71	21.6408	96	29.2608
72	21.9456	97	29.5656
73	22.2504	98	29.8704
74	22.5552	99	30.1752
75	22.8600	100	30.4800

METERS TO YARDS.

Meters.	Yards.	Meters.	Yards.
1	1.0936	26	28.4339
2	2.1872	27	29.5276
3	3.2808	28	30.6212
4	4.3745	29	31.7148
5	5.4681	30	32.8084
6	6.5617	31	33.9020
7	7.6553	32	34.9956
8	8.7489	33	36.0892
9	9.8425	34	37.1828
10	10.9361	35	38.2765
11	12.0297 .	36	39.3701
12	13.1234	37	40.4637
13	14.2170	38	41.5573
14	15.3106	39	42.6509
15	16.4042	40	43.7445
16	17.4978	41	44.8381
17	18.5914	42	45.9317
18	19.6850	43	47.0254
19	20.7786	44	48.1190
20	21.8723	45	49.2126
21	22.9659	46	50.3062
22	24.0595	47	51.3998
23	25.1531	48	52.4934
24	26.2467	49	53.5870
25	27.3403	50	54.6806

METERS TO YARDS.

Meters.	Yards.	Meters.	Yards.
51	55.7743	76	83.1146
52	56.8679	77	84.2082
53	57.9615	78	85.3018
54	59.0551	79	86.3954
55	60.1487	80	87.4890
56	61.2423	81	88.5827
57	62.3359	82	89.6763
58	63.4296	83	90.7699
59	64.5232	84	91.8635
60	65.6168	85	92.9571
61	66.7104	86	94.0507
62	67.8040	87	95.1443
63	68.8976	88	96.2379
64	69.9912	89	97.3316
65	71.0848	90	98.4252
66	72.1785	91	99.5188
67	73.2721	92	100.6124
68	74.3657	93	101.7060
69	75.4593	94	102.7996
70	76.5529	95	103.8932
71	77.6465	96	104.9868
72	78.7401	97	106.0805
73	79.8337	98	107.1741
74	80.9274	99	108.2677
75	82.0210	100	109.3613

YARDS TO METERS.

Yards.	Meters.	Yards.	Meters.
1	0.9144	26	23.7744
2	1.8288	27	24.6888
3	2.7432	28	25.6032
4	3.6576	29	26.5176
5	4.5720	30	27.4320
6	5.4864	31	28.3464
7	6.4008	32	29.2608
8	7.3152	33	30.1752
9	8.2296	34	31.0896
10	9.1440	35	32.0040
11	10.0584	36	32.9184
12	10.9728	37	33.8328
13	11.8872	38	34.7472
14	12.8016	39	35.6616
15	13.7160	40	36.5760
16	14.6304	41	37.4904
17	15.5448	42	38.4048
18	16.4592	43	39.3192
19	17.3736	44	40.2336
20	18.2880	45	41.1480
21	19.2024	46	42.0624
22	20.1168	47	42.9768
23	21.0312	48	43.8912
24	21.9456	49	44.8056
25	22.8600	50	45.7200

YARDS TO METERS.

Yards.	Meters.	Yards.	Meters.
51	46.6344	76	69.4944
52	47.5488	77	70.4088
53	48.4632	78	71.3232
54	49.3776	79	72.2376
55	50.2920	80	73.1520
56	51.2064	81	74.0664
57	52.1208	82	74.9808
58	53.0352	83	75.8952
59	53.9496	84	76.8096
60	54.8640	85	77.7240
61	55.7784	86	78.6384
62	56.6928	87	79.5528
63	57.6072	88	80.4672
64	58.5216	89	81.3816
65	59.4360	90	82.2960
66	60.3504	91	83.2104
67	61.2648	92	84.1248
68	62.1792	93	85.0392
69	63.0936	94	85.9536
70	64.0080	95	86.8680
71	64.9224	96	87.7824
72	65.8368	97	88.6968
73	66.7512	98	89.6112
74	67.6656	99	90.5256
75	68.5800	100	91.4400

KILOMETERS TO MILES.

Kilometers.	Miles.	Kilometers.	Miles.
1	0.621	26	16.156
2	1.243	27	16.777
3	1.864	28	17.398
4	2.485	29	18.020
5	3.107	30	18.641
6	3.728	31	19.262
7	4.350	32	19.884
8	4.971	33	20.505
9	5.592	34	21.127
10	6.214	35	21.748
11	6.835	36	22.369
12	7.456	37	22.991
13	8.078	38	23.612
14	8.699	39	24.233
15	9.321	40	24.855
16	9.942	41	25.476
17	10.563	42	26.098
18	11.185	43	26.719
19	11.806	44	27.340
20	12.427	45	27.962
21	13.049	46	28.583
22	13.670	47	29.204
23	14.292	48	29.826
24	14.913	49	30.447
25	15.534	50	31.069

KILOMETERS TO MILES.

Kilometers.	Miles.	Kilometers.	Miles.
51	31.690	76	47.224
52	32.311	77	47.846
53	32.933	78	48.467
54	33.554	79	49.088
55	34.175	80	49.710
56	34.797	81	50.331
57	35.418	82	50.952
58	36.040	83	51.574
59	36.661	84	52.195
60	37.282	85	52.817
61	37.904	86	53.438
62	38.525	87	54.059
63	39.146	88	54.681
64	39.768	89	55.302
65	40.389	90	55.923
66	41.010	91	56.545
67	41.632	92	57.166
68	42.253	93	57.787
69	42.875	94	58.409
70	43.496	95	59.030
71	44.117	96	59.652
72	44.739	97	60.273
73	45.360	98	60.894
74	45.981	99	61.516
75	46.603	100	61.137

MILES TO KILOMETERS.

Miles.	Kilometers.	Miles.	Kilometers.
1	1.609	26	41.843
2	3.219	27	43.452
3	4.828	28	45.062
4	6.437	29	46.671
5	8.047	30	48.280
6	9.656	31	49.890
7	11.265	32	51.499
8	12.875	33	53.108
9	14.484	34	54.718
10·	16.093	35	56.327
11	17.703	36	57.936
12	19.312	37	59.546
13	20.921	38	61.155
14	22.531	39	62.764
15	24.140	40	64.374
16	25.750	41	65.983
17	27.359	42	67.592
18	28.968	43	69.202
19	30.578	44	70.811
20	32.187	45	72.420
21	33.796	46	74.030
22	35.406	47	75.639
23	37.015	48	77.249
24	38.624	49	78.858
25	40.234	50	80.467

MILES TO KILOMETERS.

Miles.	Kilometers.	Miles.	Kilometers.
51	82.077	76	122.310
52	83.686·	77	123.919
53	85.295	78	125.529
54	86.905	79	127.138
55	88.514	80	128.748
56	90.123	81	130.357
57	91.733	82	131.966
58	93.342	83	133.576
59	94.951	84	135.185
60	96.561	85	136.794
61	98.170	86	138.404
62	99.779	87	140.013
63	101.389	88	141.622
64	102.998	89	143.232
65	104.607	90	144.841
66	106.217	91	146.450
67	107.826	92	148.060
68	109.435	93	149.669
69	111.045	94	151.278
70	112.654	95	152.888
71	114.263	96	154.497
72	115.873	97	156.106
73	117.482	98	157.716
74	119.091	99	159.325
75	120.701	100	160.934

SQUARE CENTIMETERS TO SQUARE INCHES.

Square Centimeters.	Square Inches.	Square Centimeters.	Square Inches.
1	0.15500	26	4.03001
2	0.31000	27	4.18501
3	0.46500	28	4.34001
4	0.62000	29	4.49501
5	0.77500	30	4.65001
6	0.93000	31	4.80501
7	1.08500	32	4.96001
8	1.24000	33	5.11501
9	1.39500	34	5.27001
10	1.55000	35	5.42501
11	1.70500	36	5.58001
12	1.86000	37	5.73501
13	2.01500	38	5.89001
14	2.17000	39	6.04501
15	2.32500	40	6.20001
16	2.48000	41	6.35501
17	2.63500	42	6.51001
18	2.79000	43	6.66501
19	2.94500	44	6.82001
20	3.10000	45	6.97501
21	3.25500	46	7.13001
22	3.41000	47	7.28501
23	3.56500	48	7.44001
24	3.72000	49	7.59501
25	3.87500	50	7.75001

SQUARE CENTIMETERS TO SQUARE INCHES.

Square Centimeters.	Square Inches.	Square Centimeters.	Square Inches.
51	7.90501	76	11.78002
52	8.06001	77	11.93502
53	8.21501	78	12.09002
54	8.37001	79	12.24502
55	8.52501	80	12.40002
56	8.68001	81	12.55502
57	8.83501	82	12.71002
58	8.99001	83	12.86502
59	9.14501	84	13.02002
60	9.30001	85	13.17502
61	9.45501	86	13.33002
62	9.61001	87	13.48502
63	9 76501	88	13.64002
64	9.92001	89	13.79502
65	10.07501	90	13.95002
66	10.23001	91	14.10502
67	10.38501	92	14.26002
68	10 54001	93	14.41502
69	10.69501	94	14.57002
70	10.85001	95	14.72502
71	11.00501	96	14.88002
72	11.16001	97	15.03502
73	11.31501	98	15.19002
74	11.47001	99	15.34502
75	11.62502	100	15 50002

SQUARE INCHES TO SQUARE CENTIMETERS.

Square Inches.	Square Centimeters.	Square Inches.	Square Centimeters.
1	6.452	26	167.742
2	12.903	27	174.193
3	19.355	28	180.645
4	25.806	29	187.097
5	32.258	30	193.548
6	38.710	31	200.000
7	45.161	32	206.451
8	51.613	33	212.903
9	58.064	34	219.355
10	64.516	35	225.806
11	70.968	36	232.258
12	77.419	37	238.709
13	83.871	38	245.161
14	90.322	39	251.613
15	96.774	40	258.064
16	103.226	41	264.516
17	109.677	42	270.967
18	116.129	43	277.419
19	122.581	44	283.871
20	129.032	45	290.322
21	135.484	46	296.774
22	141.935	47	303.225
23	148.387	48	309.677
24	154.839	49	316.129
25	161.290	50	322.580

SQUARE INCHES TO SQUARE CENTIMETERS.

Square Inches.	Square Centimeters.	Square Inches.	Square Centimeters.
51	329.032	76	490.322
52	335.484	77	496.774
53	341.935 ·	78	503.225
54	348.387	79	509.677
55	354.838	80	516.128
56	361.290	81	522.580
57	367.742	82	529.032 ,
58	374.193	83	535.483
59	380.645	84	541.935
60	387.096	85	548.387
61	393.548	86	554.838
62	400.000	87	561.290
63	406.451	88	567.741
64	412.903	89	574.193
65	419.354	90	580.645
66	425.806	91	587.096
67	432.258	92	593.548
68	438.709	93	599.999
69	445.161	94	606.451
70	451.612	95	612.903
71	458.064	96	619.354
72	464.516	97	625.806
73	470.967	98	632.257
74	477.419	99	638.709
75	483.870	100	645.161

SQUARE METERS TO SQUARE FEET.

Square Meters.	Square Feet.	Square Meters.	Square Feet.
1	10.764	26	279.861
2	21.528	27	290.625
3	32.292	28	301.389
4	43.056	29	312.153
5	53.820	30	322.917
6	64.583	31	333.681
7	75.347	32	344.445
8	86.111	33	355.209
9	96.875	34	365.973
10	107.639	35	376.737
11	118.403	36	387.500
12	129.167	37	398.264
13	139.931	38	409.028
14	150.695	39	419.792
15	161.459	40	430.556
16	172.222	41	441 320
17	182.986	42	452.084
18	193.750	43	462.848
19	204.514	44	473.612
20	215.278	45	484.376
21	226.042	46	495.139
22	236.806	47	505.903
23	247.570	48	516.667
24	258.334	49	527.431
25	269.098	50	538.195

SQUARE METERS TO SQUARE FEET.

Square Meters.	Square Feet.	Square Meters.	Square Feet.
51	548.959	76	818.056
52	559.723	77	828.820
53	570.487	78	839.584
54	581.251	79	850.348
55	592.015	80	861.112
56	602.778	81	871.876
57	613.542	82	882.640
58	624.306	83	893.404
59	635.070	84	904.168
60	645.834	85	914.932
61	656.598	86	925.695
62	667.362	87	936.459
63	678.126	88	947.223
64	688.890	89	957.987
65	699.654	90	968.751
66	710.417	91	979.515
67	721.181	92	990.279
68	731.945	93	1001.043
69	742.709	94	1011.807
70	753.473	95	1022.571
71	764.237	96	1033.334
72	775.001	97	1044.098
73	785.765	98	1054.862
74	796.529	99	1065.626
75	807.293	100	1076.390

SQUARE FEET TO SQUARE METERS.

Square Feet.	Square Meters	Square Feet.	Square Meters.
1	0.09290	26	2.41548
2	0.18581	27	2.50838
3	0.27871	28	2.60129
4	0.37161	29	2.69419
5	0.46452	30	2.78709
6	0.55742	31	2.88000
7	0.65032	32	2.97290
8	0.74322	33	3.06580
9	0.83613	34	3.15871
10	0.92903	35	3.25161
11	1.02193	36	3.34451
12	1.11484	37	3.43742
13	1.20774	38	3.53032
14	1.30064	39	3.62322
15	1.39355	40	3.71612
16	1.48645	41	3.80903
17	1.57935	42	3.90193
18	1.67226	43	3.99483
19	1.76516	44	4.08774
20	1.85806	45	4.18064
21	1.95097	46	4.27354
22	2.04387	47	4.36645
23	2.13677	48	4.45935
24	2.22967	49	4.55225
25	2.32258	50	4.64516

SQUARE FEET TO SQUARE METERS.

Square Feet.	Square Meters.	Square Feet.	Square Meters.
51	4.73806	76	7.06064
52	4.83096	77	7.15354
53	4.92386	78	7.24644
54	5.01677	79	7.33935
55	5.10967	80	7.43225
56	5.20257	81	7.52515
57	5.29548	82	7.61806
58	5.38838	83	7.71096
59	5.48128	84	7.80386
60	5.57419	85	7.89676
61	5.66709	86	7.98967
62	5.75999	87	8.08257
63	5.85290	88	8.17547
64	5.94580	89	8.26838
65	6.03870	90	8.36128
66	6.13161	91	8.45418
67	6.22451	92	8.54709
68	6.31741	93	8.63999
69	6.41031	94	8.73289
70	6.50322	95	8.82580
71	6 59612	96	8.91870
72	6.68902	97	9.01160
73	6.78193	98	9.10450
74	6.87483	99	9.19741
75	6.96773	100	9.29031

SQUARE METERS TO SQUARE YARDS.

Square Meters.	Square Yards.	Square Meters.	Square Yards.
I	1.1960	26	31.0957
2	2.3920	27	32.2917
3	3.5880	28	33.4877
4	4.7840	29	34.6837
5	5.9799	30	35.8797
6	7.1759	31	37.0757
7	8.3719	32	38.2716
8	9.5679	33	39.4676
9	10.7639	34	40.6636
10	11.9599	35	41.8596
11	13.1559	36	43.0556
12	14.3519	37	44.2516
13	15.5479	38	45.4476
14	16.7438	39	46.6436
15	17.9398	40	47.8396
16	19.1358	41	49.0355
17	20.3318	42	50.2315
18	21.5278	43	51.4275
19	22.7238	44	52.6235
20	23.9198	45	53.8195
21	25.1158	46	55.0155
22	26.3118	47	56.2115
23	27.5077	48	57.4075
24	28.7037	49	58.6035
25	29.8997	50	59.7994

SQUARE METERS TO SQUARE YARDS.

Square Meters.	Square Yards.	Square Meters.	Square Yards.
51	60.9954	76	90.8952
52	62.1914	77	92.0912
53	63.3874	78	93.2871
54	64.5834	79	94.4831
55	65.7794	80	95.6791
56	66.9754	81	96.8751
57	68.1714	82	98.0711
58	69.3674	83	99.2671
59	70.5634	84	100.4631
60	71.7593	85	101.6591
61	72.9553	86	102.8551
62	74.1513	87	104.0510
63	75.3473	88	105.2470
64	76.5433	89	106.4430
65	77.7393	90	107.6390
66	78.9353	91	108.8350
67	80.1313	92	110.0310
68	81.3273	93	111.2270
69	82.5232	94	112.4230
70	83.7192	95	113.6190
71	84.9152	96	114.8149
72	86.1112	97	116.0109
73	87.3072	98	117.2069
74	88.5032	99	118.4029
75	89.6992	100	119.5989

SQUARE YARDS TO SQUARE METERS.

Square Yards.	Square Meters.	Square Yards.	Square Meters.
1	0.8361	26	21.7393
2	1.6723	27	22.5755
3	2.5084	28	23.4116
4	3.3445	29	24.2477
5	4.1806	30	25.0838
6	5.0168	31	25.9200
7	5.8529	32	26.7561
8	6.6890	33	27.5922
9	7.5252	34	28.4284
10	8.3613	35	29.2645
11	9.1974	36	30.1006
12	10.0335	37	30.9367
13	10.8697	38	31.7729
14	11.7058	39	32.6090
15	12.5419	40	33.4451
16	13.3780	41	34.2813
17	14.2142	42	35.1174
18	15.0503	43	35.9535
19	15.8864	44	36.7896
20	16.7226	45	37.6258
21	17.5587	46	38.4619
22	18.3948	47	39.2980
23	19.2309	48	40.1341
24	20.0671	49	40.9703
25	20.9032	50	41.8064

SQUARE YARDS TO SQUARE METERS.

Square Yards.	Square Meters.	Square Yards.	Square Meters.
51	42.6425	76	63.5457
52	43.4787	77	64.3819
53	44.3148	78	65.2180
54	45.1509	79	66.0541
55	45.9870	80	66.8902
56	46.8232	81	67.7264
57	47.6593	82	68.5625
58	48.4954	83	69.3986
59	49.3316	84	70.2348
60	50.1677	85	71.0709
61	51.0038	86	71.9070
62	51.8399	87	72.7431
63	52.6761	88	73.5793
64	53.5122	89	74.4154
65	54.3483	90	75.2515
66	55.1845	91	76.0877
67	56.0206	92	76.9238
68	56.8567	93	77.7599
69	57.6928	94	78.5960
70	58.5290	95	79.4322
71	59.3651	96	80.2683
72	60.2012	97	81.1044
73	61.0374	98	81.9406
74	61.8735	99	82.7767
75	62.7096	100	83.6128

SQUARE KILOMETERS TO SQUARE MILES.

Square Kilometers.	Square Miles.	Square Kilometers.	Square Miles.
1	0.3861	26	10.0386
2	0.7722	27	10.4247
3	1.1583	28	10.8109
4	1.5444 .	29	11.1970
5	1.9305	30	11.5831
6	2.3166	31	11.9692
7	2.7027	32	12.3553
8	3.0888	33	12.7414
9	3.4749	34	13.1275
10	3.8610	35	13.5136
11	4.2471	36	13.8997
12	4.6332	37	14.2858
13	5.0193	38	14.6719
14	5.4054	39	15.0580
15	5.7915	40	15.4441
16	6.1776	41	15.8302
17	6.5637	42	16.2163
18	6.9498	43	16.6024
19	7.3359	44	16.9885
20	7.7220	45	17.3746
21	8.1081	46	17.7607
22	8.4942	47	18.1468
23	8.8803	48	18.5329
24	9.2664	49	18.9190
25	9.6525	50	19.3051

SQUARE KILOMETERS TO SQUARE MILES.

Square Kilometers.	Square Miles.	Square Kilometers.	Square Miles.
51	19.6912	76	29.3437
52	20.0773	77	29.7298
53	20.4634	78	30.1159
54	20.8495	79	30.5020
55	21.2356	80	30.8881
56	21.6217	81	31.2742
57	22.0078	82	31.6603
58	22.3939	83	32.0464
59	22.7800	84	32.4326
60	23.1661	85	32.8187
61	23.5522	86	33.2048
62	23.9383	87	33.5909
63	24.3244	88	33.9770
64	24.7105	89	34.3631
65	25.0966	90	34.7492
66	25.4827	91	35.1353
67	25.8688	92	35.5214
68	26.2549	93	35.9075
69	26.6410	94	36.2936
70	27.0271	95	36.6797
71	27.4132	96	37.0658
72	27.7993	97	37.4519
73	28.1854	98	37.8380
74	28.5715	99	38.2241
75	28.9576	100	38.6102

SQUARE MILES TO SQUARE KILOMETERS.

Square Miles.	Square Kilometers.	Square Miles.	Square Kilometers.
1	2.590	26	67.340
2	5.180	27	69.930
3	7.770	28	72.520
4	10.360	29	75.110
5	12.950	30	77.700
6	15.540	31	80.290
7	18.130	32	82.880
8	20.720	33	85.470
9	23.310	34	88.060
10	25.900	35	90.650
11	28.490	36	93.240
12	31.080	37	95.830
13	33.670	38	98.420
14	36.260	39	101.010
15	38.850	40	103.600
16	41.440	41	106.190
17	44.030	42	108.780
18	46.620	43	111.370
19	49.210	44	113.960
20	51.800	45	116.550
21	54.390	46	119.139
22	56.980	47	121.729
23	59.570	48	124.319
24	62.160	49	126.909
25	64.750	50	129.499

SQUARE MILES TO SQUARE KILOMETERS.

Square Miles.	Square Kilometers.	Square Miles.	Square Kilometers.
51	132.089	76	196.839
52	134.679	77	199.429
53	137.269	78	202.019
54	139.859	79	204.609
55	142.449	80	207.199
56	145.039	81	209.789
57	147.629	82	212.379
58	150.219	83	214.969
59	152.809	84	217.559
60	155.399	85	220.149
61	157.989	86	222.739
62	160.579	87	225.329
63	163.169	88	227 919
64	165.759	89	230.509
65	168.349	90	233.099
66	170.939	91	235.689
67	173.529	92	238.279
68	176.119	93	240.869
69	178.709	94	243.459
70	181.299	95	246.049
71	183.889	96	248.639
72	186.479	97	251.229
73	189.069	98	253.819
74	191.659	99	256.409
75	194.249	100	258.999

HECTARES TO ACRES.

Hectares.	Acres.	Hectares.	Acres.
1	2.471	26	64.247
2	4.942	27	66.718
3	7.413	28	69.189
4	9.884	29	71.661
5	12.355	30	74.132
6	14.826	31	76.603
7	17.297	32	79.074
8	19.768	33	81.545
9	22.239	34	84.016
10	24.711	35	86.487
11	27.182	36	88.958
12	29.653	37	91.429
13	32.124	38	93.900
14	34.595	39	96.371
15	37.066	40	98.842
16	39.537	41	101.313
17	42.008	42	103.784
18	44.479	43	106.255
19	46.950	44	108.726
20	49.421	45	111.197
21	51.892	46	113.668
22	54.363	47	116.139
23	56.834	48	118.610
24	59.305	49	121.082
25	61.776	50	123.553

HECTARES TO ACRES.

Hectares.	Acres.	Hectares.	Acres.
51	126.024	76	187.800
52	128.495	77	190.271
53	130.966	78	192.742
54	133.437	79	195.213
55	135.908	80	197.684
56	138.379	81	200.155
57	140.850	82	202.626
58	143.321	83	205.097
59	145.792	84	207.568
60	148.263	85	210.039
61	150.734	86	212.510
62	153.205	87	214.982
63	155.676	88	217.453
64	158.147	89	219.924
65	160.618	90	222.395
66	163.089	91	224.866
67	165.560	92	227.337
68	168.032	93	229.808
69	170.503	94	232.279
70	172.974	95	234.750
71	175.445	96	237.221
72	177.916	97	239.692
73	180.387	98	242.163
74	182.858	99	244.634
75	185.329	100	247.105

ACRES TO HECTARES.

Acres.	Hectares.	Acres.	Hectares.
1	0.4047	26	10.5218
2	0.8094	27	10.9265
3	1.2141	28	11.3312
4	1.6187	29	11.7359
5	2.0234	30	12.1406
6	2.4281	31	12.5453
7	2.8328	32	12.9499
8	3.2375	33	13.3546
9	3.6422	34	13.7593
10	4.0469	35	14.1640
11	4.4515	36	14.5687
12	4.8562	37	14.9734
13	5.2609	38	15.3781
14	5.6656	39	15.7827
15	6.0703	40	16.1874
16	6.4750	41	16.5921
17	6.8797	42	16.9968
18	7.2843	43	17.4015
19	7.6890	44	17.8062
20	8.0937	45	18.2109
21	8.4984	46	18.6155
22	8.9031	47	19.0202
23	9.3078	48	19.4249
24	9.7125	49	19.8296
25	10.1171	50	20.2343

ACRES TO HECTARES.

Acres.	Hectares.	Acres.	Hectares.
51	20.6390	76	30.7561
52	21.0437	77	31 1608
53	21.4483	78	31.5655
54	21.8530	79	31.9702
55	22.2577	80	32.3749
56	22.6624	81	32.7795
57	23.0671	82	33.1842
58	23.4718	83	33.5889
59	23.8765	84	33.9936
60	24.2811	85	34.3983
61	24.6858	86	34.8030
62	25.0905	87	35.2077
63	25.4952	88	35.6124
64	25.8999	89	36.0170
65	26.3046	90	36.4217
66	26.7093	91	36.8264
67	27.1139	92	37.2311
68	27.5186	93	37.6358
69	27.9233	94	38.0405
70	28.3280	95	38.4452
71	28.7327	96	38.8498
72	29.1374	97	39.2545
73	29.5421	98	39.6592
74	29.9467	99	40.0639
75	30.3514	100	40.4686

CUBIC CENTIMETERS TO CUBIC INCHES.

Cubic Centimeters.	Cubic Inches.	Cubic Centimeters.	Cubic Inches.
1	0.06102	26	1.58662
2	0.12205	27	1.64764
3	0.18307	28	1.70866
4	0.24409	29	1.76969
5	0.30512	30	1.83071
6	0.36614	31	1.89173
7	0.42717	32	1.95276
8	0.48819	33	2.01378
9	0.54921	34	2.07480
10	0.61024	35	2.13583
11	0.67126	36	2.19685
12	0.73228	37	2.25788
13	0.79331	38	2.31890
14	0.85433	39	2.37992
15	0.91535	40	2.44095
16	0.97638	41	2.50197
17	1.03740	42	2.56299
18	1.09843	43	2.62402
19	1.15945	44	2.68504
20	1.22047	45	2.74606
21	1.28150	46	2.80709
22	1.34252	47	2.86811
23	1.40354	48	2.92914
24	1.46457	49	2.99016
25	1.52559	50	3.05118

CUBIC CENTIMETERS TO CUBIC INCHES.

Cubic Centimeters.	Cubic Inches.	Cubic Centimeters.	Cubic Inches.
51	3.11221	76	4.63780
52	3.17323	77	4.69882
53	3.23425	78	4.75985
54	3.29528	79	4.82087
55	3.35630	80	4.88189
56	3.41732	81	4.94292
57	3.47835	82	5.00394
58	3.53937	83	5.06496
59	3.60040	84	5.12599
60	3.66142	85	5.18701
61	3.72244	86	5.24803
62	3.78347	87	5.30906
63	3.84449	88	5.37008
64	3.90551	89	5.43111
65	3.96654	90	5.49213
66	4.02756	91	5.55315
67	4.08859	92	5.61418
68	4.14961	93	5.67520
69	4.21063	94	5.73622
70	4.27166	95	5.79725
71	4.33268	96	5.85827
72	4.39370	97	5.91930
73	4.45473	98	5.98032
74	4.51575	99	6.04134
75	4.57677	100	6.10237

CUBIC INCHES TO CUBIC CENTIMETERS.

Cubic Inches.	Cubic Centimeters.	Cubic Inches.	Cubic Centimeters.
1	16.387	26	426.064
2	32.774	27	442.451
3	49.161	28	458.838
4	65.548	29	475 225
5	81.935	30	491.612
6	98.322	31	508.000
7	114.710	32	524.387
8	131.097	33	540.774
9	147.484	34	557.161
10	163.871	35	573.548
11	180.258	36	589.935
12	196.645	37	606.322
13	213.032	38	622.709
14	229.419	39	639.096
15	245.806	40	655.483
16	262.193	41	671.870
17	278.580	42	688.257
18	294.967	43	704.645
19	311.355	44	721.032
20	327.742	45	737.419
21	344.129	46	753.806
22	360.516	47	770.193
23	376.903	48	786.580
24	393.290	49	802.967
25	409.677	50	819.354

CUBIC INCHES TO CUBIC CENTIMETERS.

Cubic Inches.	Cubic Centimeters.	Cubic Inches.	Cubic Centimeters.
51	835.741	76	1245.418
52	852.128	77	1261.805
53	868.515	78	1278.192
54	884.902	79	1294.580
55	901.290	80	1310.967
56	917.677	81	1327.354
57	934.064	82	1343.741
58	950.451	83	1360.128
59	966.838	84	1376.515
60	983.225	85	1392.902
61	999.612	86	1409.289
62	1015.999	87	1425.676
63	1032.386	88	1442.063
64	1048.773	89	1458.450
65	1065.160	90	1474.837
66	1081.547	91	1491.225
67	1097.935	92	1507.612
68	1114.322	93	1523.999
69	1130.709	94	1540.386
70	1147.096	95	1556.773
71	1163.483	96	1573.160
72	1179.870	97	1589.547
73	1196.257	98	1605.934
74	1212.644	99	1622.321
75	1229.031	100	1638.708

CUBIC METERS TO CUBIC FEET.

Cubic Meters.	Cubic Feet.	Cubic Meters.	Cubic Feet.
1	35.31	26	918.18
2	70.63	27	953.50
3	105.94	28	988.81
4	141.26	29	1024.12
5	176.57	30	1059.44
6	211.89	31	1094.75
7	247.20	32	1130.07
8	282.52	33	1165.38
9	317.83	34	1200.70
10	353.15	35	1236.01
11	388.46	36	1271.33
12	423.78	37	1306.64
13	459.09	38	1341.96
14	494.40	39	1377.27
15	529.72	40	1412.59
16	565.03	41	1447.90
17	600.35	42	1483.21
18	635.66	43	1518 53
19	670.98	44	1553.84
20	706.29	45	1589.16
21	741.61	46	1624.47
22	776.92	47	1659.79
23	812.24	48	1695.10
24	847.55	49	1730.42
25	882.87	50	1765.73

CUBIC METERS TO CUBIC FEET.

Cubic Meters.	Cubic Feet.	Cubic Meters.	Cubic Feet.
51	1801.05	76	2683.91
52	1836.36	77	2719.23
53	1871.68	78	2754.54
54	1906.99	79	2789.86
55	1942.30	80	2825.17
56	1977.62	81	2860.49
57	2012.93	82	2895.80
58	2048.25	83	2931.11
59	2083.56	84	2966.43
60	2118.88	85	3001.74
61	2154.19	86	3037.06
62	2189.51	87	3072.37
63	2224.82	88	3107.69
64	2260.14	89	3143.00
65	2295.45	90	3178.32
66	2330.77	91	3213.63
67	2366.08	92	3248.95
68	2401.39	93	3284.26
69	2436.71	94	3319.58
70	2472.02	95	3354.89
71	2507.34	96	3390.20
72	2542.65	97	3425.52
73	2577.97	98	3460.83
74	2613.28	99	3496.15
75	2648.60	100	3531.46

CUBIC FEET TO CUBIC METERS.

Cubic Feet.	Cubic Meters.	Cubic Feet.	Cubic Meters.
1	0.02832	26	0.73624
2	0.05663	27	0.76456
3	0.08495	28	0.79287
4	0.11327	29	0.82119
5	0.14158	30	0.84951
6	0.16990	31	0.87782
7	0.19822	32	0.90614
8	0.22654	33	0.93446
9	0.25485	34	0.96277
10	0.28317	35	0.99109
11	0.31149	36	1.01941
12	0.33980	37	1.04772
13	0.36812	38	1.07604
14	0.39644	39	1.10436
15	0.42475	40	1.13268
16	0.45307	41	1.16099
17	0.48139	42	1.18931
18	0.50970	43	1.21763
19	0.53802	44	1.24594
20	0.56634	45	1.27426
21	0.59465	46	1.30258
22	0.62297	47	1.33089
23	0.65129	48	1.35921
24	0.67961	49	1.38753
25	0.70792	50	1.41584

CUBIC FEET TO CUBIC METERS.

Cubic Feet.	Cubic Meters.	Cubic Feet.	Cubic Meters.
51	1.44416	76	2.15208
52	1.47248	77	2.18040
53	1.50079	78	2.20872
54	1.52911	79	2.23703
55	1.55743	80	2.26535
56	1.58575	81	2.29367
57	1.61406	82	2.32198
58	1.64238	83	2.35030
59	1.67070	84	2.37862
60	1.69901	85	2.40693
61	1.72733	86	2.43525
62	1.75565	87	2.46357
63	1.78396	88	2.49189
64	1.81228	89	2.52020
65	1.84060	90	2.54852
66	1.86891	91	2.57684
67	1.89723	92	2.60515
68	1.92555	93	2.63347
69	1.95386	94	2.66179
70	1.98218	95	2.69010
71	2.01050	96	2.71842
72	2.03882	97	2.74674
73	2.06713	98	2.77505
74	2.09545	99	2.80337
75	2.12377	100	2.83169

CUBIC METERS TO CUBIC YARDS.

Cubic Meters.	Cubic Yards.	Cubic Meters.	Cubic Yards.
1	1.308	26	34.007
2	2.616	27	35.315
3	3.924	28	36.623
4	5.232	29	37.931
5	6.540	30	39.238
6	7.848	31	40.546
7	9.156	32	41.854
8	10.464	33	43.162
9	11.772	34	44.470
10	13.079	35	45.778
11	14.387	36	47.086
12	15.695	37	48.394
13	17.003	38	49.702
14	18.311	39	51.010
15	19.616	40	52.318
16	20.927	41	53.626
17	22.235	42	54.934
18	23.543	43	56.242
19	24.851	44	57.550
20	26.159	45	58.858
21	27.467	46	60.166
22	28.775	47	61.474
23	30.083	48	62.782
24	31.391	49	64.090
25	32.699	50	65.397

CUBIC METERS TO CUBIC YARDS.

Cubic Meters.	Cubic Yards.	Cubic Meters.	Cubic Yards.
51	66.705	76	99.404
52	68.013	77	100.712
53	69.321	78	102.020
54	70.629	79	103.328
55	71.937	80	104.636
56	73.245	81	105.944
57	74.553	82	107.252
58	75.861	83	108.560
59	77.169	84	109.868
60	78.477	85	111.176
61	79.785	86	112.484
62	81.093	87	113.792
63	82.401	88	115.100
64	83.709	89	116.407
65	85.017	90	117.715
66	86.325	91	119.023
67	87.633	92	120.331
68	88.941	93	121.639
69	90.248	94	122.947
70	91.556	95	124.255
71	92.864	96	125.563
72	94.172	97	126.871
73	95.480	98	128.179
74	96.788	99	129.487
75	98.096	100	130.795

CUBIC YARDS TO CUBIC METERS.

Cubic Yards.	Cubic Meters.	Cubic Yards.	Cubic Meters.
1	0.765	26	19.879
2	1.529	27	20.643
3	2.294	28	21.408
4	3.058	29	22.172
5	3.823	30	22.937
6	4.587	31	23.701
7	5.352	32	24.466
8	6.116	33	25.230
9	6.881	34	25.995
10	7.646	35	26.760
11	8.410	36	27.524
12	9.175	37	28.289
13	9.939	38	29.053
14	10.704	39	29.818
15	11.468	40	30.582
16	12.233	41	31.347
17	12.998	42	32.112
18	13.762	43	32.876
19	14.527	44	33.641
20	15.291	45	34.405
21	16.056	46	35.170
22	16.820	47	35.934
23	17.585	48	36.699
24	18.349	49	37.463
25	19.114	50	38.228

CUBIC YARDS TO CUBIC METERS.

Cubic Yards.	Cubic Meters.	Cubic Yards.	Cubic Meters.
51	38.993	76	58.107
52	39.757	77	58.871
53	40.522	78	59.636
54	41.286	79	60.400
55	42.051	80	61.165
56	42.815	81	61.929
57	43.580	82	62.694
58	44.344	83	63.458
59	45.109	84	64.223
60	45.874	85	64.988
61	46.638	86	65.752
62	47.403	87	66.517
63	48.167	88	67.281
64	48.932	89	68.046
65	49.696	90	68.810
66	50.461	91	69.575
67	51.226	92	70.340
68	51.990	93	71.104
69	52.755	94	71.869
70	53.519	95	72.633
71	54.284	96	73.398
72	55.048	97	74.162
73	55.813	98	74.927
74	56.577	99	75.691
75	57.342	100	76.456

· CUBIC KILOMETERS TO CUBIC MILES.

Cubic Kilometers.	Cubic Miles.	Cubic Kilometers.	Cubic Miles.
1	0.2399	26	6.2377
2	0.4798	27	6.4776 .
3	0.7197	28	6.7175
4	0.9596	29	6.9575
5	1.1996	30	7.1974
6	1.4395	31	7.4373
7	1.6794	32	7.6772
8	1.9193	33	7.9171
9	2.1592	34	8.1570
10	2.3991	35	8.3969
11	2.6390	36	8.6368
12	2.8789	37	8.8768
13	3.1189	38	9.1167
14	3.3588	39	9.3566
15	3.5987	40	9.5965
16	3.8386	41	9.8364
17	4.0785	42	10.0763
18	4.3184	43	10.3162
19	4.5583	44	10.5561
20	4.7982	45	10.7961
21	5.0382	46	11.0360
22	5.2781	47	11.2759
23	5.5180	48	11.5158
24	5.7579	49	11.7557
25	5 9978	50	11.9956

CUBIC KILOMETERS TO CUBIC MILES.

Cubic Kilometers.	Cubic Miles.	Cubic Kilometers.	Cubic Miles..
51	12.2355	76	18.2333
52	12.4754	77	18.4733
53	12.7154	78	18.7132
54	12.9553	79	18.9531
55	13.1952	80	19.1930
56	13.4351	81	19.4329
57	13.6750	82	19.6728
58	13.9149	83	19.9127
59	14.1548	84	20.1526
60	14.3947	85	20.3926
61	14.6347	86	20.6325
62	14.8746	87	20.8724
63	15.1145	88	21.1123
64	15.3544	89	21.3522
65	15.5943	90	21.5921
66	15.8342	91	21.8320
67	16.0741	92	22.0719
68	16.3140	93	22.3119
69	16.5540	94	22.5518
70	16.7939	95	22.7917
71	17.0338	96	23.0316
72	17.2737	97	23.2715
73	17.5136	98	23.5114
74	17.7535	99	23.7513
75	17.9934	100	23.9912

CUBIC MILES TO CUBIC KILOMETERS.

Cubic Miles.	Cubic Kilometers.	Cubic Miles.	Cubic Kilometers.
1	4.168	26	108.373
2	8.336	27	112.541
3	12.505	28	116.709
4	16.673	29	120.877
5	20.841	30	125.046
6	25.009	31	129.214
7	29.177	32	133.382
8	33.345	33	137.550
9	37.514	34	141.718
10	41.682	35	145.886
11	45.850	36	150.055
12	50.018	37	154.223
13	54.186	38	158.391
14	58.355	39	162.559
15	62.523	40	166.727
16	66.691	41	170.896
17	70.859	42	175.064
18	75.027	43	179.232
19	79.196	44	183.400
20	83.364	45	187.568
21	87.532	46	191.737
22	91.700	47	195.905
23	95.868	48	200.073
24	100.036	49	204.241
25	104.205	50	208.409

CUBIC MILES TO CUBIC KILOMETERS.

Cubic Miles.	Cubic Kilometers.	Cubic Miles.	Cubic Kilometers.
51	212.577	76	316.782
52	216.746	77	320.950
53	220.914	78	325.118
54	225.082	79	329.287
55	229.250	80	333.455
56	233.418	81	337.623
57	237.587	82	341.791
58	241.755	83	345.959
59	245.923	84	350.128
60	250.091	85	354.296
61	254.259	86	358.464
62	258.428	87	362.632
63	262.596	88	366.800
64	266.764	89	370.969
65	270.932	90	375.137
66	275.100	91	379.305
67	279.268	92	383.473
68	283.437	93	387.641
69	287.605	94	391.809
70	291.773	95	395.978
71	295.941	96	400.146
72	300.109	97	404.314
73	304.278	98	408.482
74	308.446	99	412.650
75	312.614	100	416.819

CUBIC CENTIMETERS TO U. S. FLUID OUNCES.

Cubic Centimeters.	Fluid Ounces.	Cubic Centimeters.	Fluid Ounces.
1	0.0338	26	0.8789
2	0.0676	27	0.9127
3	0.1014	28	0.9465
4	0.1352	29	0.9803
5	0.1690	30	1.0141
6	0.2028	31	1.0479
7	0.2366	32	1.0817
8	0.2704	33	1.1155
9	0.3042	34	1.1493
10	0.3380	35	1.1831
11	0.3718	36	1.2169
12	0.4056	37	1.2507
13	0.4394	38	1.2845
14	0.4732	39	1.3183
15	0.5070	40	1.3521
16	0.5408	41	1.3859
17	0.5746	42	1.4197
18	0.6084	43	1.4535
19	0.6422	44	1.4873
20	0.6760	45	1.5211
21	0.7098	46	1.5549
22	0.7436	47	1.5887
23	0.7775	48	1.6225
24	0.8113	49	1.6563
25	0.8451	50	1.6901

CUBIC CENTIMETERS TO U. S. FLUID OUNCES.

Cubic Centimeters.	Fluid Ounces.	Cubic Centimeters.	Fluid Ounces.
51	1.7239	76	2.5690
52	1.7577	77	2.6028
53	1.7915	78	2.6366
54	1.8253	79	2.6704
55	1.8591	80	2.7042
56	1.8929	81	2.7380
57	1.9267	82	2.7718
58	1.9605	83	2.8056
59	1.9943	84	2.8394
60	2.0281	85	2.8732
61	2.0619	86	2.9070
62	2.0957	87	2.9408
63	2.1295	88	2.9746
64	2.1633	89	3.0084
65	2.1971	90	3.0422
66	2.2309	91	3.0760
67	2.2647	92	3.1098
68	2.2985	93	3.1436
69	2.3324	94	3.1774
70	2.3662	95	3.2112
71	2.4000	96	3.2450
72	2.4338	97	3.2788
73	2.4676	98	3.3126
74	2.5014	99	3.3464
75	2.5352	100	3.3802

U. S. FLUID OUNCES TO CUBIC CENTIMETERS.

Fluid Ounces.	Cubic Centimeters.	Fluid Ounces.	Cubic Centimeters.
1	29.6	26	769.2
2	59.2	27	798.8
3	88.8	28	828.3
4	118.3	29	857.9
5	147.9	30	887.5
6	177.5	31	917.1
7	207.1	32	946.7
8	236.7	33	976.3
9	266.3	34	1005.9
10	295.8	35	1035.4
11	325.4	36	1065.0
12	355.0	37	1094.6
13	384.6	38	1124.2
14	414.2	39	1153.8
15	443.8	40	1183.4
16	473.3	41	1212.9
17	502.9	42	1242.5
18	532.5	43	1272.1
19	562.1	44	1301.7
20	591.7	45	1331.3
21	621.3	46	1360.9
22	650.8	47	1390.4
23	680.4	48	1420.0
24	710.0	49	1449.6
25	739.6	50	1479.2

U. S. FLUID OUNCES TO CUBIC CENTIMETERS.

Fluid Ounces.	Cubic Centimeters.	Fluid Ounces.	Cubic Centimeters.
51	1508.8	76	2248.4
52	1538.4	77	2278.0
53	1567.9	78	2307.5
54	1597.5	79	2337.1
55	1627.1	80	2366.7
56	1656.7	81	2396.3
57	1686.3	82	2425.9
58	1715.9	83	2455.5
59	1745.5	84	2485.0
60	1775.0	85	2514.6
61	1804.6	86	2544.2
62	1834.2	87	2573.8
63	1863.8	88	2603.4
64	1893.4	89	2633.0
65	1923.0	90	2662.6
66	1952.5	91	2692.1
67	1982.1	92	2721.7
68	2011.7	93	2751.3
69	2041.3	94	2780.9
70	2070.9	95	2810.5
71	2100.5	96	2840.1
72	2130.0	97	2869.6
73	2159.6	98	2899.2
74	2189.2	99	2928.8
75	2218.8	100	2958.4

LITERS TO U. S. LIQUID QUARTS.

Liters.	Quarts.	Liters.	Quarts.
1	1.056	26	27.464
2	2.113	27	28.520
3	3.169	28	29.577
4	4.225	29	30.633
5	5.282	30	31.689
6	6.338	31	32.745
7	7.394	32	33.802
8	8.450	33	34.858
9	9.507	34	35.914
10	10.563	35	36.971
11	11.619	36	38.027
12	12.676	37	39.083
13	13.732	38	40.140
14	14.788	39	41.196
15	15.844	40	42.252
16	16.901	41	43.308
17	17.957	42	44.365
18	19.013	43	45.421
19	20.070	44	46.477
20	21.126	45	47.534
21	22.182	46	48.590
22	23.239	47	49.646
23	24.295	48	50.703
24	25.351	49	51.759
25	26.408	50	52.815

LITERS TO U. S. LIQUID QUARTS.

Liters.	Quarts.	Liters.	Quarts.
51	53.872	76	80.279
52	54.928	77	81.335
53	55.984	78	82.392
54	57.040	79	83.448
55	58.097	80	84.504
56	59.153	81	85.561
57	60 209	82	86.617
58	61.266	83	87.673
59	62.322	84	88.730
60	63.378	85	89.786
61	64.435	86	90.842
62	65.491	87	91.898
63	66.547	88	92.955
64	67.603	89	94.011
65	68.660	90	95.067
66	69.716	91	96.124
67	70.772	92	97.180
68	71.829	93	98.236
69	72.885	94	99.293
70	73.941	95	100.349
71	74.998	96	101.405
72	76.054	97	102.461
73	77.110	98	103.518
74	78.166	99	104.574
75	79.223	100	105.630

U. S. LIQUID QUARTS TO LITERS.

Quarts.	Liters.	Quarts.	Liters.
1	0.947	26	24.614
2	1.893	27	25.561
3	2.840	28	26.508
4	3.787	29	27.454
5	4.733	30	28.401
6	5.680	31	29.348
7	6.627	32	30.294
8	7.574	33	31.241
9	8.520	34	32.188
10	9.467	35	33.134
11	10.414	36	34.081
12	11.360	37	35.028
13	12.307	38	35.975
14	13.254	39	36.921
15	14.200	40	37.868
16	15.147	41	38.815
17	16.094	42	39.761
18	17.041	43	40.708
19	17.987	44	41.655
20	18.934	45	42.601
21	19.881	46	43.548
22	20.827	47	44.495
23	21.774	48	45.442
24	22.721	49	46.388
25	23.667	50	47.335

U. S. LIQUID QUARTS TO LITERS.

Quarts.	Liters.	Quarts.	Liters.
51	48.282	76	71.949
52	49.228	77	72.896
53	50.175	78	73.842
54	51.122	79	74.789
55	52.068	80	75.736
56	53.015	81	76.683
57	53 962	82	77.629
58	54.908	83	78 576
59	55.855	84	79.523
60	56.802	85	80.469
61	57.749	86	81.416
62	58.695	87	82.363
63	59.642	88	83.309
64	60.589	89	84.256
65	61.535	90	85.203
66	62.482	91	86.150
67	63.429	92	87.096
68	64.375	93	88.043
69	65.322	94	88.990
70	66.269	95	89.936
71	67.216	96	90.883
72	68.162	97	91.830
73	69.109	98	92.776
74	70.056	99	93.723
75	71.002	100	94.670

LITERS TO U. S. GALLONS.

Liters.	U. S. Gallons.	Liters.	U. S. Gallons.
1	0.264	26	6.866
2	0.528	27	7.130
3	0.792	28	7.394
4	1.056	29	7 658
5	1.320	30	7.922
6	1.584	31	8.186
7	1.849	32	8.451
8	2.113	33	8 715
9	2.377	34	8 979
10	2.641	35	9.243
11	2.905	36	9.507
12	3.169	37	9.771
13	3.433	38	10 035
14	3.697	39	10.299
15	3.961	40	10.563
16	4.225	41	10.827
17	4.489	42	11.091
18	4.753	43	11.355
19	5.018	44	11.620
20	5.282	45	11.884
21	5.546	46	12.148
22	5.810	47	12.412
23	6.074	48	12.676
24	6.338	49	12.940
25	6.602	50	13.204

LITERS TO U. S. GALLONS.

Liters.	U. S. Gallons.	Liters.	U. S. Gallons.
51	13.468	76	20.070
52	13.732	77	20.334
53	13.996	78	20.598
54	14.260	79	20.862
55	14.524	80	21.126
56	14.788	81	21.390
57	15.053	82	21.655
58	15.317	83	21.919
59	15.581	84	22.183
60	15.845	85	22.447
61	16.109	86	22.711
62	16.373	87	22.975
63	16.637	88	23.239
64	16.901	89	23.503
65	17.165	90	23.767
66	17.429	91	24.031
67	17.693	92	24.295
68	17.957	93	24.559
69	18.222	94	24.824
70	18.486	95	25.088
71	18.750	96	25.352
72	19.014	97	25.616
73	19.278	98	25.880
74	19.542	99	26.144
75	19.8c6	100	26.408

U. S. GALLONS TO LITERS.

U. S. Gallons.	Liters.	U. S. Gallons.	Liters.
1	3.79	26	98.46
2	7.57	27	102.24
3	11.36	28	106.03
4	15.15	29	109.82
5	18.93	30	113.60
6	22.72	31	117.39
7	26.51	32	121.18
8	30.29	33	124.96
9	34.08	34	128.75
10	37.87	35	132.54
11	41.65	36	136.32
12	45.44	37	140.11
13	49.23	38	143.90
14	53.02	39	147.68
15	56.80	40	151.47
16	60.59	41	155.26
17	64.38	42	159.05
18	68.16	43	162.83
19	71.95	44	166.62
20	75.74	45	170.41
21	79.52	46	174.19
22	83.31	47	177.98
23	87.10	48	181.77
24	90.88	49	185.55
25	94.67	50	189.34

U. S. GALLONS TO LITERS.

U. S. Gallons.	Liters.	U. S. Gallons.	Liters.
51	193.13	76	287.80
52	196.91	77	291.58
53	200.70	78	295.37
54	204.49	79	299.16
55	208.27	80	302.94
56	212.06	81	306.73
57	215.85	82	310.52
58	219.63	83	314.30
59	223.42	84	318.09
60	227.21	85	321.88
61	230.99	86	325.66
62	234.78	87	329.45
63	238.57	88	333.24
64	242.35	89	337.02
65	246.14	90	340.81
66	249.93	91	344.60
67	253.71	92	348.38
68	257.50	93	352.17
69	261.29	94	355.96
70	265.08	95	359.75
71	268.86	96	363.53
72	272.65	97	367.32
73	276.44	98	371.11
74	280.22	99	374.89
75	284.01	100	378.68

LITERS TO BRITISH LIQUID QUARTS.

Liters.	British Quarts.	Liters.	British Quarts.
1	0.88	26	22.90
2	1.76	27	23.78
3	2.64	28	24.67
4	3.52	29	25.55
5	4 40	30	26 43
6	5.29	31	27.31
7	6.17	32	28.19
8	7.05	33	29.07
9	7.93	34	29 95
10	8.81	35	30.83
11	9.69	36	31.71
12	10.57	37	32.59
13	11.45	38	33.47
14	12.33	39	34.36
15	13.21	40	35.24
16	14.09	41	36.12
17	14.98	42	37.00
18	15.86	43	37.88
19	16.74	44	38.76
20	17.62	45	39.64
21	18.50	46	40.52
22	19.38	47	41.40
23	20.26	48	42.28
24	21.14	49	43.16
25	22.02	50	44.04

LITERS TO BRITISH LIQUID QUARTS.

Liters.	British Quarts.	Liters.	British Quarts.
51	44.93	76	66.95
52	45.81	77	67.83
53	46.69	78	68.71
54	47.57	79	69.59
55	48.45	80	70.47
56	49.33	81	71.35
57	50.21	82	72.23
58	51.09	83	73.11
59	51.97	84	74.00
60	52.85	85	74.88
61	53.73	86	75.76
62	54.62	87	76.64
63	55.50	88	77.52
64	56.38	89	78.40
65	57.26	90	79.28
66	58.14	91	80.16
67	59.02	92	81.04
68	59.90	93	81.92
69	60.78	94	82.80
70	61.66	95	83.69
71	62.54	96	84.57
72	63.42	97	85.45
73	64.31	98	86.33
74	65.19	99	87.21
75	66.07	100	88.09

BRITISH LIQUID QUARTS TO LITERS.

British Quarts.	Liters.	British Quarts.	Liters.
1	1.14	26	29.52
2	2.27	27	30.65
3	3.41	28	31.79
4	4.54	29	32.92
5	5.68	30	34.06
6	6.81	31	35.19
7	7.95	32	36.33
8	9.08	33	37.46
9	10.22	34	38.60
10	11.35	35	39.73
11	12.49	36	40.87
12	13.62	37	42.00
13	14.76	38	43.14
14	15.89	39	44.27
15	17.03	40	45.41
16	18.16	41	46.55
17	19.30	42	47.68
18	20.43	43	48.82
19	21.57	44	49.95
20	22.71	45	51.09
21	23.84	46	52.22
22	24.98	47	53.36
23	26.11	48	54.49
24	27.25	49	55.63
25	28.38	50	56.76

BRITISH LIQUID QUARTS TO LITERS.

British Quarts.	Liters.	British Quarts.	Liters.
51	57.90	76	86.28
52	59.03	77	87.41
53	60.17	78	88.55
54	61.30	79	89.69
55	62.44	80	90.82
56	63.57	81	91.96
57	64.71	82	93.09
58	65.84	83	94.23
59	66.98	84	95.36
60	68.12	85	96.50
61	69.25	86	97.63
62	70.39	87	98.77
63	71.52	88	99.90
64	72.66	89	101.04
65	73.79	90	102.17
66	74.93	91	103.31
67	76.06	92	104.44
68	77.20	93	105.58
69	78.33	94	106.71
70	79.47	95	107.85
71	80.60	96	108.98
72	81.74	97	110.12
73	82.87	98	111.25
74	84.01	99	112.39
75	85.14	100	113.53

LITERS TO BRITISH GALLONS.

Liters.	British Gallons.	Liters.	British Gallons.
1	0.220	26	5.726
2	0.440	27	5.946
3	0.661	28	6.166
4	0.881	29	6.386
5	1.101	30	6.607
6	1.321	31	6.827
7	1.542	32	7.047
8	1.762	33	7.267
9	1.982	34	7.487
10	2.202	35	7.708
11	2.422	36	7.928
12	2.643	37	8.148
13	2.863	38	8.368
14	3.083	39	8.589
15	3.303	40	8.809
16	3.524	41	9.029
17	3.744	42	9.249
18	3.964	43	9.469
19	4.184	44	9.690
20	4.404	45	9.910
21	4.625	46	10.130
22	4.845	47	10.350
23	5.065	48	10.571
24	5.285	49	10.791
25	5.505	50	11.011

LITERS TO BRITISH GALLONS.

Liters.	British Gallons.	Liters.	British Gallons.
51	11.231	76	16.737
52	11.451	77	16.957
53	11.672	78	17.177
54	11.892	79	17.397
55	12.112	80	17.618
56	12.332	81	17.838
57	12.553	82	18.058
58	12.773	83	18.278
59	12.993	84	18.498
60	13.213	85	18.719
61	13.433	86	18.939
62	13.654	87	19.159
63	13.874	88	19.379
64	14.094	89	19.600
65	14.314	90	19.820
66	14.535	91	20.040
67	14.755	92	20.260
68	14.975	93	.20.480
69	15.195	94	20.701
70	15.415	95	20.921
71	15.636	96	21.141
72	15.856	97	21.361
73	16.076	98	21.582
74	16.296	99	21.802
75	16.516	100	22.022

BRITISH GALLONS TO LITERS.

British Gallons.	Liters.	British Gallons.	Liters
1	4.54	26	118.07
2	9.08	27	122.61
3	13 62	28	127.15
4	18.16	29	131.69
5	22.71	30	136.23
6	27.25	31	140.77
7	31.79	32	145.31
8	36.33	33	149.85
9	40.87	34	154.39
10	45.41	35	158.94
11	49.95	36	163.48
12	54.49	37	168.02
13	59.03	38	172.56
14	63.57	39	177.10
15	68.12	40	181.64
16	72.66	41	186.18
17	77.20	42	190.72
18	81.74	43	195.26
19	86.28	44	199.80
20	90.82	45	204.35
21	95.36	46	208.89
22	99.90	47	213.43
23	104 44	48	217.97
24	108.98	49	222.51
25	113.53	50	227.05

BRITISH GALLONS TO LITERS.

British Gallons.	Liters.	British Gallons.	Liters.
51	231.59	76	345.12
52	236.13	77	349.66
53	240.67	78	354.20
54	245.21	79	358.74
55	249.76	80	363.28
56	254.30	81	367.82
57	258.84	82	372.36
58	263.38	83	376.90
59	267.92	84	381.45
60	272.46	85	385.99
61	277.00	86	390.53
62	281.54	87	395.07
63	286 08	88	399.61
64	290.63	89	404.15
65	295.17	90	408.69
66	299.71	91	413.23
67	304.25	92	417.77
68	308.79	93	422.31
69	313.33	94	426.86
70	317.87	95	431.40
71	322.41	96	435.94
72	326.95	97	440.48
73	331.49	98	445.02
74	336.04	99	449 56
75	340.58	100	454.10

LITERS TO U. S. DRY QUARTS.

Liters.	Dry Quarts.	Liters.	Dry Quarts.
1	0.91	26	23.61
2	1.82	27	24.52
3	2.72	28	25.43
4	3.63	29	26.33
5	4.54	30	27.24
6	5.45	31	28.15
7	6.36	32	29 06
8	7.26	33	29.97
9	8.17	34	30.87
10	9.08	35	31.78
11	9.99	36	32.69
12	10.90	37	33.60
13	11.81	38	34 51
14	12.71	39	35.42
15	13.62	40	36.32
16	14.53	41	37.23
17	15.44	42	38.14
18	16.35	43	39.05
19	17.25	44	39 96
20	18.16	45	40.86
21	19 07	46	41 77
22	19.93	47	42.68
23	20 89	48	43.59
24	21.79	49	44.50
25	22.70	50	45.40

LITERS TO U. S. DRY QUARTS.

Liters.	Dry Quarts.	Liters.	Dry Quarts.
51	46.31	76	69.01
52	47.22	77	69.92
53	48.13	78	70.83
54	49.04	79	71.74
55	49.94	80	72.65
56	50.85	81	73.55
57	51.76	82	74.46
58	52.67	83	75.37
59	53.58	84	76.28
60	54.48	85	77.19
61	55.39	86	78.09
62	56.30	87	79.00
63	57.21	88	79.91
64	58.12	89	80.82
65	59.03	90	81.73
66	59.93	91	82.64
67	60.84	92	83.54
68	61.75	93	84.45
69	62.66	94	85.36
70	63.57	95	86.27
71	64.47	96	87.18
72	65.38	97	88.08
73	66.29	98	88.99
74	67.20	99	89.90
75	68.11	100	90.81

U. S. DRY QUARTS TO LITERS.

Dry Quarts.	Liters.	Dry Quarts.	Liters.
1	1.10	26	28.63
2	2.20	27	29.73
3	3.30	28	30.83
4	4.40	29	31.93
5	5.51	30	33.04
6	6.61	31	34.14
7	7.71	32	35.24
8	8.81	33	36.34
9	9.91	34	37.44
10	11.01	35	38.54
11	12.11	36	39.64
12	13.21	37	40.74
13	14.32	38	41.85
14	15.42	39	42.95
15	16.52	40	44.05
16	17.62	41	45.15
17	18.72	42	46.25
18	19.82	43	47.35
19	20.92	44	48.45
20	22.02	45	49.55
21	23.13	46	50.66
22	24.23	47	51.76
23	25.33	48	52 86
24	26 43	49	53.96
25	27.53	50	55.06

U. S. DRY QUARTS TO LITERS.

Dry Quarts.	Liters.	Dry Quarts.	Liters.
51	56.16	76	83.69
52	57.26	77	84.79
53	58.36	78	85.89
54	59.46	79	86.99
55	60.57	80	88.10
56	61.67	81	89.20
57	62.77	82	90.30
58	63.87	83	91.40
59	64.97	84	92.50
60	66.07	85	93.60
61	67.17	86	94.70
62	68.27	87	95.80
63	69.38	88	96.91
64	70.48	89	98.01
65	71.58	90	99.11
66	72.68	91	100.21
67	73.78	92	101 31
68	74.88	93	102.41
69	75 98	94	103.51
70	77.08	95	104.61
71	78.19	96	105.72
72	79.29	97	106.82
73	80.39	98	107.92
74	81.49	99	109.02
75	82.59	100	110.12

CUBIC METERS (Steres) TO U. S. BUSHELS.

Cubic Meters (Steres.)	Bushels.	Cubic Meters (Steres.)	Bushels.
1	28.4	26	737.8
2	56.8	27	766.2
3	85.1	28	794.6
4	113.5	29	823.0
5	141.9	30	851.3
6	170.3	31	879.7
7	198.6	32	908.1
8	227.0	33	936.5
9	255.4	34	964.8
10	283.8	35	993.2
11	312.2	36	1021.6
12	340.5	37	1050.0
13	368.9	38	1078.3
14	397.3	39	1106.7
15	425.7	40	1135.1
16	454.0	41	1163.5
17	482.4	42	1191.9
18	510.8	43	1220.2
19	539.2	44	1248.6
20	567.6	45	1277.0
21	595.9	46	1305.4
22	624.3	47	1333.7
23	652.7	48	1362.1
24	681.1	49	1390.5
25	709.4	50	1418.9

CUBIC METERS (Steres) TO U. S. BUSHELS.

Cubic Meters (Steres.)	Bushels.	Cubic Meters (Steres.)	Bushels.
51	1447.3	76	2156.7
52	1475.6	77	2185.1
53	1504.0	78	2213.5
54	1532.4	79	2241.8
55	1560.8	80	2270.2
56	1589.1	81	2298.6
57	1617.5	82	2327 0
58	1645 9	83	2355.3
59	1674.3	84	2383.7
60	1702.7	85	2412.1
61	1731.0	86	2440.5
62	1759.4	87	2468.9
63	1787 8	88	2497.2
64	1816.2	89	2525.6
65	1844.5	90	2554.0
66	1872 9	91	2582.4
67	1901.3	92	2610.7
68	1929.7	93	2639.1
69	1958.1	94	2667.5
70	1986.4	95	2695.9
71	2014 8	96	2724.2
72	2043.2	97	2752.6
73	2071.6	98	2781.0
74	2099 9	99	2809.4
75	2128.3	100	2837.8

U. S. BUSHELS TO CUBIC METERS (Steres.)

Bushels.	Cubic Meters (Steres.)	Bushels.	Cubic Meters (Steres.)
1	0.0352	26	0.9162
2	0.0705	27	0 9515
3	0.1057	28	0.9867
4	0.1410	29	1.0219
5	0.1762	30	1.0572
6	0.2114	31	1.0924
7	0.2467	32	1.1276
8	0.2819	33	1.1629
9	0 3172	34	1.1981
10	0.3524	35	1.2334
11	0.3876	36	1.2686
12	0.4229	37	1.3038
13	0 4581	38	1.3391
14	0.4933	39	1.3743
15	0.5286	40	1.4096
16	0.5638	41	1.4448
17	0.5991	42	1.4800
18	0.6343	43	1.5153
19	0.6695	44	1.5505
20	0.7048	45	1.5858
21	0 7400	46	1.6210
22	0.7753	47	1.6562
23	0.8105	48	1.6915
24	0.8457	49	1.7267
25	0.8810	50	1.7620

U. S. BUSHELS TO CUBIC METERS (Steres.)

Bushels.	Cubic Meters (Steres.)	Bushels.	Cubic Meters (Steres.)
51	1.7972	76	2.6782
52	1.8324	77	2.7134
53	1.8677	78	2.7486
54	1.9029	79	2.7839
55	1.9381	80	2.8191
56	1.9734	81	2.8544
57	2.0086	82	2.8896
58	2.0439	83	2.9248
59	2.0791	84	2.9601
60	2.1143	85	2.9953
61	2.1496	86	3.0306
62	2.1848	87	3.0658
63	2.2201	88	3.1010
64	2.2553	89	3.1363
65	2.2905	90	3.1715
66	2.3258	91	3.2067
67	2.3610	92	3.2420
68	2.3963	93	3.2772
69	2.4315	94	3.3125
70	2.4667	95	3.3477
71	2.5020	96	3.3829
72	2.5372	97	3.4182
73	2.5724	98	3.4534
74	2.6077	99	3.4887
75	2.6429	100	3.5239

GRAMMES TO GRAINS.

Grammes.	Grains.	Grammes.	Grains.
1	15.432	26	401.241
2	30.865	27	416.673
3	46 297	28	432.106
4	61.729	29	447.538
5	77 162	30	462 970
6	92.594	31	478.403
7	108.026	32	493.835
8	123.459	33	509.268
9	138.891	34	524.700
10	154.323	35	540.132
11	169.756	36	555.565
12	185.188	37	570.997
13	200.621	38	586.429
14	216.053	39	601.862
15	231.485	40	617.294
16	246.918	41	632.726
17 .	262.350	42	648.159
18	277.782	43	663.591
19	293.215	44	679.023
20	308.647	45	694.456
21	324.079	46	709.888
22	339 512	47	725.320
23	354.944	48	740.753
24	370.376	49	756.185
25	385.809	50	771.617

GRAMMES TO GRAINS.

Grammes.	Grains.	Grammes.	Grains.
51	787.050	76	1172 859
52	802 482	77	1188.291
53	817.914	78	1203.723
54	833.347	79	1219.156
55	848.779	80	1234.588
56	864.212	81	1250.020
57	879.644	82	1265.453
58	895.076	83	1280.885
59	910.509	84	1296.317
60	925.941	85	1311.750
61	941.373	86	1327.182
62	956.806	87	1342.614
63	972.238	88	1358.047
64	987.670	89	1373.479
65	1003.103	90	1388.911
66	1018.535	91	1404.344
67	1033.967	92	1419.776
68	1049.400	93	1435.208
69	1064.832	94	1450.641
70	1080.264	95	1466.073
71	1095.697	96	1481.505
72	1111.129	97	1496.938
73	1126.561	98	1512.370
74	1141.994	99	1527.803
75	1157.426	100	1543.235

GRAINS TO GRAMMES.

Grains.	Grammes.	Grains.	Grammes.
1	0.0648	26	1.6848
2	0.1296	27	1.7496
3	0.1944	28	1.8144
4	0.2592	29	1.8792
5	0.3240	30	1.9440
6	0.3888	31	2.0088
7	0.4536	32	2.0736
8	0.5184	33	2.1384
9	0.5832	34	2.2032
10	0.6480	35	2.2680
11	0 7128	36	2.3328
12	0 7776	37	2 3976
13	0 8424	38	2.4624
14	0.9072	39	2.5272
15	0.9720	40	2.5920
16	1.0368	41	2.6568
17	1.1016	42	2.7216
18	1.1664	43	2.7864
19	1.2312	44	2.8512
20	1.2960	45	2.9160
21	1.3608	46	2.9808
22	1.4256	47	3.0456
23	1.4904	48	3.1104
24	1.5552	49	3.1752
25	1.6200	50	3.2400

GRAINS TO GRAMMES.

Grains.	Grammes.	Grains.	Grammes.
51	3.3047	76	4 9247
52	3.3695	77	4.9895
53	3.4343	78	5.0543
54	3.4991	79	5.1191
55	3 5639	80	5.1839
56	3.6287	81	5.2487
57	3.6935	82	5.3135
58	3.7583	83	5.3783
59	3.8231	84	5.4431
60	3 8879	85	5 5079
61	3 9527	86	5.5727
62	4 0175	87	5.6375
63	4.0823	88	5.7023
64	4.1471	89	5.7671
65	4.2119	90	5.8319
66	4.2767	91	5.8967
67	4.3415	92	5.9615
68	4.4063	93	6.0263
69	4.4711	94	6.0911
70	4.5359	95	6.1559
71	4.6007	96	6.2207
72	4.6655	97	6.2855
73	4.7303	98	6.3503
74	4.7951	99	6.4151
75	4.8599	100	6.4799

GRAMMES TO AVOIRDUPOIS OUNCES.

Grammes.	Ounces.	Grammes.	Ounces.
1	0.035274	26	0.917122
2	0.070548	27	0.952396
3	0.105822	28	0 987670
4	0.141096	29	1.022944
5	0.176370	30	1.058218
6	0.211644	31	1 093492
7	0.246918	32	1.128766
8	0.282192	33	1.164040
9	0.317465	34	1.199314
10	0.352739	35	1.234588
11	0.388013	36	1.269862
12	0.423287	37	1.305136
13	0 458561	38	1.340410
14	0 493835	39	1.375684
15	0.529109	40	1.410958
16	0.564383	41	1 446232
17	0.599657	42	1.481505
18	0.634931	43	1.516779
19	0.670203	44	1.552053
20	0.705479	45	1.587327
21	0.740753	46	1.622601
22	0.776027	47	1.657875
23	0.811301	48	1.693149
24	0.846575	49	1.728423
25	0.881848	50	1.763697

GRAMMES TO AVOIRDUPOIS OUNCES.

Grammes.	Ounces.	Grammes.	Ounces.
51	1.798971	76	2.680819
52	1.834245	77	2.716093
53	1.869519	78	2.751367
54	1.904793	79	2.786641
55	1.940067	80	2.821915
56	1.975341	81	2.857189
57	2.010615	82	2.892463
58	2.045889	83	2 927737
59	2.081162	84	2.963011
60	2.116436	85	2.998285
61	2.151710	86	3.033559
62	2.186984	87	3.068833
63	2.222258	88	3.104107
64	2.257532	89	3.139381
65	2.292806	90	3.174655
66	2.328080	91	3.209929
67	2.363354	92	3.245202
68	2.398628	93	3.280476
69	2.433902	94	3.315750
70	2.469176	95	3.351024
71	2.504450	96	3.386298
72	2.539724	97	3 421572
73	2.574998	98	3.456846
74	2.610272	99	3.492120
75	2.645545	100	3.527394

AVOIRDUPOIS OUNCES TO GRAMMES.

Ounces.	Grammes.	Ounces.	Grammes.
1	28.350	26	737.088
2	56.699	27	765.438
3	85.049	28	793.787
4	113.398	29	822.137
5	141.748	30	850.486
6	170.097	31	878.836
7	198.447	32	907.185
8	226.796	33	935.535
9	255.146	34	963.884
10	283.495	35	992.234
11	311.845	36	1020.583
12	340.194	37	1048.933
13	368.544	38	1077.283
14	396.894	39	1105.632
15	425.243	40	1133.982
16	453.593	41	1162.331
17	481.942	42	1190.681
18	510.292	43	1219.030
19	538.641	44	1247.380
20	566.991	45	1275.729
21	595.340	46	1304.079
22	623.690	47	1332.428
23	652.039	48	1360.778
24	680.389	49	1389.127
25	708.738	50	1417.477

AVOIRDUPOIS OUNCES TO GRAMMES.

Ounces.	Grammes.	Ounces.	Grammes.
51	1445.827	76	2154.565
52	1474.176	77	2182.915
53	1502.526	78	2211.264
54	1530.875	79	2239.614
55	1559.225	80	2267.963
56	1587.574	81	2296.313
57	1615.924	82	2324.662
58	1644.273	83	2353.012
59	1672.623	84	2381.361
60	1700.972	85	2409.711
61	1729.322	86	2438.060
62	1757.671	87	2466.410
63	1786.021	88	2494.760
64	1814.371	89	2523.109
65	1842 720	90	2551.459
66	1871.070	91	2579.808
67	1899.419	92	2608.158
68	1927.769	93	2636.507
69	1956.118	94	2664.857
70	1984.468	95	2693.206
71	2012.817	96	2721.556
72	2041.167	97	2749 905
73	2069.516	98	2778.255
74	2097.866	99	2806.604
75	2126.215	100	2834.954

KILOGRAMS TO AVOIRDUPOIS POUNDS.

Kilograms.	Pounds.	Kilograms.	Pounds.
1	2.205	26	57.320
2	4.409	27	59.525
3	6.614	28	61.729
4	8.818	29	63.934
5	11.023	30	66.139
6	13.228	31	68.343
7	15.432	32	70.548
8	17.637	33	72.752
9	19.842	34	74.957
10	22.046	35	77.162
11	24.251	36	79.366
12	26.455	37	81.571
13	28.660	38	83.776
14	30.865	39	85.980
15	33.069	40	88.185
16	35.274	41	90.389
17	37.479	42	92.594
18	39.683	43	94.799
19	41.888	44	97.003
20	44.092	45	99.208
21	46.297	46	101.413
22	48.502	47	103.617
23	50.706	48	105.822
24	52.911	49	108.026
25	55.116	50	110.231

KILOGRAMS TO AVOIRDUPOIS POUNDS.

Kilograms.	Pounds.	Kilograms.	Pounds.
51	112.436	76	167.551
52	114 640	77	169.756
53	116.845	78	171.960
54	119.050	79	174.165
55	121.254	80	176.370
56	123 459	81	178.574
57	125.663	82	180.779
58	127.868	83	182.984
59	130 073	84	185.188
60	132.277	85	187.393
61	134.482	86	189.597
62	136.687	87	191.802
63	138.891	88	194.007
64	141 096	89	196.211
65	143 300	90	198.416
66	145.505	91	200.621
67	147.710	92	202.825
68	149.914	93	205.030
69	152.119	94	207 234
70	154.323	95	209.439
71	156.528	96	211.644
72	158.733	97	213.848
73	160.937	98	216.053
74	163.142	99	218.257
75	165.347	100	220.462

AVOIRDUPOIS POUNDS TO KILOGRAMS.

Pounds.	Kilograms.	Pounds.	Kilograms.
1	0.4536	26	11.7934
2	0.9072	27	12.2470
3	1.3608	28	12.7006
4	1.8144	29	13.1542
5	2.2680	30	13.6078
6	2.7216	31	14.0614
7	3.1752	32	14.5150
8	3.6287	33	14.9686
9	4.0823	34	15.4222
10	4.5359	35	15.8758
11	4.9895	36	16.3293
12	5.4431	37	16.7829
13	5.8967	38	17.2365
14	6.3503	39	17.6901
15	6.8039	40	18.1437
16	7.2575	41	18.5973
17	7.7111	42	19.0509
18	8.1647	43	19.5045
19	8.6183	44	19.9581
20	9.0719	45	20.4117
21	9.5255	46	20.8653
22	9.9790	47	21.3189
23	10.4326	48	21.7725
24	10.8862	49	22.2261
25	11.3398	50	22.6797

AVOIRDUPOIS POUNDS TO KILOGRAMS.

Pounds.	Kilograms.	Pounds.	Kilograms.
51	23.1332	76	34.4731
52	23.5868	77	34.9267
53	24.0404	78	35.3803
54	24.4940	79	35.8338
55	24.9476	80	36.2874
56	25.4012	81	36.7410
57	25.8548	82	37.1946
58	26.3084	83	37.6482
59	26.7620	84	38.1018
60	27.2156	85	38.5554
61	27.6692	86	39 0090
62	28.1228	87	39.4626
63	28.5764	88	39.9162
64	29.0300	89	40.3698
65	29.4835	90	40.8234
66	29.9371	91	41.2770
67	30.3907	92	41.7306
68	30.8443	93	42.1841
69	31.2979	94	42.6377
70	31.7515	95	43.0913
71	32.2051	96	43.5449
72	32.6587	97	43.9985
73	33.1123	98	44.4521
74	33.5659	99	44.9057
75	34.0195	100	45.3593

GRAMMES TO TROY OUNCES.

Grammes.	Ounces.	Grammes.	Ounces.
1	0.0322	26	0.8359
2	0.0643	27	0.8681
3	0.0965	28	0.9002
4	0.1286	29	0.9324
5	0.1608	30	0.9645
6	0.1929	31	0 9967
7	0.2251	32	1.0288
8	0.2572	33	1.0610
9	0.2894	34	1.0931
10	0.3215	35	1.1253
11	0.3537	36	1.1574
12	0.3858	37	1.1896
13	0.4180	38	1.2217
14	0.4501	39	1.2539
15	0.4823	40	1.2860
16	0.5144	41	1.3182
17	0.5466	42	1.3503
18	0.5787	43	1.3825
19	0.6109	44	1.4146
20	0.6430	45	1.4468
21	0.6752	46	1.4789
22	0.7073	47	1.5111
23	0.7395	48	1.5432
24	0.7716	49	1.5754
25	0.8038	50	1.6075

GRAMMES TO TROY OUNCES.

Grammes.	Ounces.	Grammes.	Ounces.
51	1.6397	76	2 4435
52	1.6718	77	2.4756
53	1.7040	78	2.5078
54	1.7361	79	2 5399
55	1 7683	80	2.5721
56	1.8004	81	2.6042
57	1.8326	82	2.6364
58	1.8647	83	2 6685
59	1.8969	84	2.7007
60	1.9290	85	2.7328
61	1.9612	86	2.7650
62	1.9933	87	2.7971
63	2.0255	88	2 8293
64	2.0576	89	2.8614
65	2.0898	90	2.8936
66	2.1219	91	2 9257
67	2.1541	92	2.9579
68	2.1862	93	2.9900
69	2.2184	94	3.0222
70	2.2505	95	3.0543
71	2.2827	96	3 0865
72	2.3149	97	3 1186
73	2 3470	98	3.1508
74	2 3792	99	3.1829
75	2.4113	100	3.2151

TROY OUNCES TO GRAMMES.

Ounces.	Grammes.	Ounces.	Grammes.
1	31.1	26	808.7
2	62.2	27	839.8
3	93.3	28	870 9
4	124.4	29	902.0
5	155.5	30	933.1
6	186.6	31	964.2
7	217.7	32	995.3
8	248.8	33	1026.4
9	279.9	34	1057.5
10	311.0	35	1088.6
11	342.1	36	1119.7
12	373.2	37	1150.8
13	404.4	38	1182.0
14	435.5	39	1213.1
15	466.6	40	1244.2
16	497.7	41	1275.3
17	528.8	42	1306.4
18	559.9	43	1337.5
19	591.0	44	1368.6
20	622.1	45	1399.7
21	653.2	46	1430.8
22	684.3	47	1461.9
23	715.4	48	1493.0
24	746.5	49	1524.1
25	777.6	50	1555.2

TROY OUNCES TO GRAMMES.

Ounces.	Grammes.	Ounces.	Grammes.
51	1586.3	76	2363 9
52	1617.4	77	2395.0
53	1648.5	78	2426.1
54	1679 6	79	2457.2
55	1710.7	80	2488.3
56	1741.8	81	2519 4
57	1772.9	82	2550.5
58	1804.0	83	2581.6
59	1835.1	84	2612.7
60	1866.2	85	2643.8
61	1897.3	86	2674.9
62	1928.4	87	2706.0
63	1959.6	88	2737.2
64	1990 7	89	2768 3
65	2021.8	90	2799.4
66	2052.9	91	2830.5
67	2084.0	92	2861 6
68	2115.1	93	2892 7
69	2146.2	94	2923 8
70	2177.3	95	2954.9
71	2208 4	96 ·	2986.0
72	2239.5	97	3017.1
73	2270.6	98	3048.2
74	2301.7	99	3079.3
75	2332.8	100	3110.4

KILOGRAMS TO TROY POUNDS.

Kilograms.	Pounds.	Kilograms.	Pounds.
1	2.68	26	69.66
2	5.36	27	72.34
3	8.04	28	75.02
4	10.72	29	77.70
5	13.40	30	80.38
6	16 08	31	83.06
7	18.75	32	85.74
8	21.43	33	88.41
9	24.11	34	91.09
10	26.79	35	93 77
11	29.47	36	96.45
12	32.15	37	99 13
13	34.83	38	101.81
14	37.51	39	104.49
15	40.19	40	107.17
16	42.87	41	109.85
17	45.55	42	112.53
18	48.23	43	115.21
19	50.91	44	117.89
20	53.58	45	120.57
21	56.26	46	123.24
22	58.94	47	125.92
23	61.62	48	128.60
24	64.30	49	131.28
25	66.98	50	133.96

KILOGRAMS TO TROY POUNDS.

Kilograms.	Pounds.	Kilograms.	Pounds.
51	136.64	76	203.62
52	139.32	77	206.30
53	142.00	78	208.98
54	144.68	79	211.66
55	147.36	80	214.34
56	150 04	81	217.02
57	152.72	82	219.70
58	155.40	83	222.38
59	158.07	84	225.06
60	160.75	85	227.73
61	163.43	86	230.41
62	166.11	87	233.09
63	168.79	88	235.77
64	171.47	89	238.45
65	174.15	90	241.13
66	176.83	91	243.81
67	179.51	92	246.49
68	182.19	93	249.17
69	184.87	94	251.85
70	187.55	95	254.53
71	190.23	96	257.21
72	192.90	97	259.89
73	195.58	98	262.56
74	198.26	99	265.24
75	200.94	100	267.92

TROY POUNDS TO KILOGRAMS.

Pounds.	Kilograms.	Pounds.	Kilograms.
1	0.373	26	9.704
2	0.746	27	10.077
3	1.120	28	10.451
4	1.493	29	10.824
5	1.866	30	11.197
6	2.239	31	11.570
7	2.613	32	11.944
8	2.986	33	12.317
9	3.359	34	12.690
10	3.732	35	13.063
11	4.106	36	13.437
12	4.479	37	13.810
13	4.852	38	14.183
14	5.225	39	14.556
15	5.599	40	14.930
16	5.972	41	15.303
17	6.345	42	15.676
18	6.718	43	16.049
19	7.092	44	16.423
20	7.465	45	16.796
21	7.838	46	17.169
22	8.211	47	17.542
23	8.585	48	17.916
24	8.958	49	18.289
25	9.331	50	18.662

TROY POUNDS TO KILOGRAMS.

Pounds.	Kilograms.	Pounds.	Kilograms.
51	19.035	76	28.366
52	19.408	77	28.739
53	19.782	78	29.113
54	20.155	79	29.486
55	20.528	80	29.859
56	20.901	81	30.232
57	21.275	82	30.606
58	21.648	83	30.979
59	22.021	84	31.352
60	22.394	85	31.725
61	22.768	86	32.099
62	23.141	87	32.472
63	23 514	88	32.845
64	23.887	89	33.218
65	24.261	90	33.592
66	24.634	91	33.965
67	25.007	92	34.338
68	25.380	93	34.711
69	25.754	94	35.085
70	26.127	95	35.458
71	26.500	96	35.831
72	26.873	97	36.204
73	27.247	98	36.578
74	27 620	99	36.951
75	27.993	100	37.324

MILLIERS TO SHORT TONS.

Milliers.	Tons (of 2000 lbs.)	Milliers.	Tons (of 2000 lbs.)
1	1.102	26	28.660
2	2.205	27	29 762
3	3.307	28	30.865
4	4.409	29	31.967
5	5.512	30	33.069
6	6.614	31	34.172
7	7.716	32	35.274
8	8.818	33	36.376
9	9.921	34	37.479
10	11.023	35	38.581
11	12.125	36	39.683
12	13.228	37	40.785
13	14.330	38	41.888
14	15.432	39	42 990
15	16 535	40	44.092
16	17.637	41	45.195
17	18.739	42	46.297
18	19.842	43	47.399
19	20.944	44	48.502
20	22.046	45	49.604
21	23.149	46	50.706
22	24.251	47	51.809
23	25.353	48	52.911
24	26.455	49	54.013
25	27.558	50	55.116

MILLIERS TO SHORT TONS.

Milliers.	Tons (of 2000 lbs.)	Milliers.	Tons (of 2000 lbs.)
51	56.218	76	83.776
52	57.320	77	84.878
53	58.422	78	85.980
54	59.525	79	87.083
55	60.627	80	88.185
56	61.729	81	89.287
57	62.832	82	90.389
58	63.934	83	91.492
59	65.036	84	92.594
60	66.139	85	93.696
61	67.241	86	94.799
62	68.343	87	95.901
63	69.446	88	97.003
64	70.548	89	98.106
65	71.650	90	99.208
66	72.753	91	100.310
67	73.855	92	101.413
68	74.957	93	102.515
69	76.059	94	103.617
70	77.162	95	104.720
71	78.264	96	105.822
72	79.366	97	106.924
73	80.469	98	108.026
74	81.571	99	109.129
75	82.673	100	110.231

SHORT TONS TO MILLIERS.

Tons (of 2000 lbs.)	Milliers.	Tons (of 2000 lbs.)	Milliers.
1	0.907	26	23.587
2	1.814	27	24.494
3	2.722	28	25.401
4	3.629	29	26.309
5	4.536	30	27.216
6	5.443	31	28.123
7	6.350	32	29.030
8	7.258	33	29.937
9	8.165	34	30.844
10	9.072	35	31.752
11	9.979	36	32.659
12	10.886	37	33.566
13	11.793	38	34.473
14	12.701	39	35.380
15	13.608	40	36.288
16	14.515	41	37.195
17	15.422	42	38.102
18	16.329	43	39.009
19	17.237	44	39.916
20	18.144	45	40.824
21	19.051	46	41.731
22	19.958	47	42.638
23	20.865	48	43.545
24	21.773	49	44.452
25	22.680	50	45.360

SHORT TONS TO MILLIERS.

Tons (of 2000 lbs.)	Milliers.	Tons (of 2000 lbs.)	Milliers.
51	46.267	76	68.946
52	47.174	77	69.854
53	48.081	78	70.761
54	48.988	79	71.668
55	49.895	80	72.575
56	50.803	81	73.482
57	51.710	82	74.390
58	52.617	83	75.297
59	53.524	84	76.204
60	54.431	85	77.111
61	55.339	86	78.018
62	56.246	87	78.926
63	57.153	88	79.833
64	58.060	89	80.740
65	58.967	90	81.647
66	59.875	91	82.554
67	60.782	92	83.461
68	61.689	93	84.369
69	62.596	94	85.276
70	63.503	95	86.183
71	64.410	96	87.090
72	65.318	97	87.997
73	66.225	98	88.905
74	67.132	99	89.812
75	68.039	100	90.719

KILOGRAMS PER SQUARE METER TO POUNDS PER SQUARE FOOT.

Kilograms per Square Meter.	Pounds per Square Foot.	Kilograms per Square Meter	Pounds per Square Foot.
1	0 205	26	5.325
2	0.410	27	5.530
3	0.614	28	5.735
4	0.819	29	5.940
5	1.024	30	6.144
6	1.229	31	6.349
7	1.434	32	6.554
8	1.639	33	6.759
9	1.843	34	6.964
10	2.048	35	7.169
11	2.253	36	7.373
12	2.458	37	7.578
13	2.663	38	7.783
14	2.867	39	7.988
15	3.072	40	8.193
16	3.277	41	8.397
17	3.482	42	8.602
18	3.687	43	8.807
19	3.892	44	9.012
20	4.096	45	9.217
21	4.301	46	9.422
22	4.506	47	9.626
23	4.711	48	9.831
24	4.916	49	10.036
25	5.120	50	10.241

KILOGRAMS PER SQUARE METER TO POUNDS PER SQUARE FOOT.

Kilograms per Square Meter.	Pounds per Square Foot.	Kilograms per Square Meter.	Pounds per Square Foot.
51	10.446	76	15.566
52	10.650	77	15.771
53	10.855	78	15.976
54	11.060	79	16.180
55	11.265	80	16.385
56	11.470	81	16.590
57	11.675 ·	82	16.795
58	11.879	83	17.000
59	12.084	84	17.205
60	12.289	85	17.409
61	12.494	86	17.614
62	12.699	87	17.819
63	12.903	88	18.024
64	13.108	89	18.229
65	13.313	90	18.433
66	13.518	91	18.638
67	13.723	92	18.843
68	13.927	93	19.048
69	14.132	94	19.253
70	14.337	95	19.458
71	14.542	96	19.662
72	14.747	97	19 867
73	14.952	98	20.072
74	15.156	99	20.277
75	15.361	100	20.482 ·

POUNDS PER SQUARE FOOT TO KILOGRAMS PER SQUARE METER.

Pounds per Square Foot.	Kilograms per Square Meter.	Pounds per Square Foot.	Kilograms per Square Meter.
1	4.88	26	126.94
2	9.76	27	131.82
3	14.65	28	136.71
4	19.53	29	141.59
5	24.41	30	146.47
6	29.29	31	151.35
7	34.18	32	156.24
8	39.06	33	161.12
9	43 94	34	166.00
10	48.82	35	170.88
11	53.71	36	175.77
12	58.59	37	180.65
13	63.47	38	185.53
14	68.35	39	190.41
15	73.24	40	195.30
16	78.12	41	200 18
17	83.00	42	205.06
18	87.88	43	209.94
19	92.77	44	214.83
20	97.65	45	219.71
21	102.53	46	224.59
22	107.41	47	229.47
23	112.30	48	234.36
24	117.18	49	239.24
25	122.06	50	244.12

POUNDS PER SQUARE FOOT TO KILOGRAMS PER SQUARE METER.

Pounds per Square Foot.	Kilograms per Square Meter.	Pounds per Square Foot.	Kilograms per Square Meter.
51	249.00	76	371.06
52	253.88	77	375.94
53	258.77	78	380.83
54	263.65	79	385.71
55	268.53	80	390.59
56	273.41	81	395.47
57	278.30	82	400.36
58	283.18	83	405.24
59	288.06	84	410.12
60	292.94	85	415.00
61	297.83	86	419.89
62	302.71	87	424.77
63	307.59	88	429.65
64	312.47	89	434.53
65	317.36	90	439.42
66	322.24	91	444.30
67	327.12	92	449.18
68	332.00	93	454.06
69	336.89	94	458.95
70	341.77	95	463.83
71	346.65	96	468.71
72	351.53	97	473.59
73	356.42	98	478.48
74	361.30	99	483.36
75	366.18	100	488.24

KILOGRAMS PER SQUARE CENTIMETER TO POUNDS PER SQUARE INCH.

Kilograms per Square Centimeter.	Pounds per Square Inch.	Kilograms per Square Centimeter.	Pounds per Square Inch.
1	14.22	26	369.81
2	28.45	27	384.03
3	42.67	28	398.25
4	56.89	29	412.48
5	71.12	30	426.70
6	85.34	31	440 92
7	99.56	32	455.15
8	113.79	33	469.37
9	128.01	34	483 59
10	142.23	35	497.82
11	156.46	36	512.04
12	170.68	37	526.26
13	184.90	38	540.49
14	199 13	39	554.71
15	213 35	40	568.93
16	227.57	41	583.16
17	241.80	42	597.38
18	256.02	43	611.60
19	270.24	44	625.83
20	284.47	45	640.05
21	298.69	46	654.27
22	312 91	47	668.50
23	327.14	48	682.72
24	341.36	49	696.94
25	355.58	50	711.17

KILOGRAMS PER SQUARE CENTIMETER TO POUNDS PER SQUARE INCH.

Kilograms per Square Centimeter.	Pounds per Square Inch.	Kilograms per Square Centimeter.	Pounds per Square Inch.
51	725.39	76	1080 97
52	739.61	77	1095.20
53	753.84	78	1109.42
54	768.06	79	1123.64
55	782.28	80	1137.87
56	796.51	81	1152.09
57	810.73	82	1166.31
58	824.95	83	1180 54
59	839 18	84	1194 76
60	853.40	85	1208.98
61	867.62	86	1223.21
62	881.85	87	1237.43
63	896.07	88	1251.65
64	910 29	89	1265.88
65	924.52	90	1280.10
66	938.74	91	1294.32
67	952.96	92	1308 55
68	967.19	93	1322.77
69	981.41	94	1336 99
70	995.63	95	1351.22
71	1009 86	96	1365.44
72	1024.08	97	1379.66
73	1038.30	98	1393 89
74	1052.53	99	1408.11
75	1066.75	100	1422.34

POUNDS PER SQUARE INCH TO KILOGRAMS PER SQUARE CENTIMETER.

Pounds per Square Inch.	Kilograms per Square Centimeter.	Pounds per Square Inch.	Kilograms per Square Centimeter.
1	0.0703	26	1 8280
2	0.1406	27	1.8983
3	0.2109	28	1.9686
4	0.2812	29	2.0389
5	0.3515	30	2.1092
6	0.4218	31	2.1795
7	0.4921	32	2.2498
8	0.5625	33	2.3201
9	0.6328	34	2.3904
10	0.7031	35	2.4607
11	0.7734	36	2.5311
12	0.8437	37	2.6014
13	0 9140	38	2.6717
14	0.9843	39	2.7420
15	1.0546	40	2.8123
16	1.1249	41	2.8826
17	1.1952	42	2.9529
18	1.2655	43	3.0232
19	1.3358	44	3.0935
20	1.4061	45	3.1638
21	1.4764	46	3.2341
22	1.5468	47	3.3044
23	1.6171	48	3.3747
24	1.6874	49	3.4450
25	1.7577	50	3.5154

POUNDS PER SQUARE INCH TO KILOGRAMS PER SQUARE CENTIMETER.

Pounds per Square Inch.	Kilograms per Square Centimeter	Pounds per Square Inch.	Kilograms per Square Centimeter.
51	3.5857	76	5.3433
52	3.6560	77	5.4136
53	3.7263	78	5.4839
54	3.7966	79	5.5543
55	3.8669	80	5.6246
56	3.9372	81	5.6949
57	4.0075	82	5.7652
58	4.0778	83	5.8355
59	4.1481	84	5.9058
60	4.2184	85	5.9761
61	4.2887	86	6.0464
62	4.3590	87	6.1167
63	4.4293	88	6.1870
64	4 4996	89	6.2573
65	4.5700	90	6.3276
66	4.6403	91	6.3979
67	4.7106	92	6.4682
68	4.7809	93	6.5386
69	4.8512	94	6.6089
70	4.9215	95	6.6792
71	4.9918	96	6.7495
72	5.0621	97	6.8198
73	5.1324	98	6.8901
74	5.2027	99	6.9604
75	5.2730	100	7.0307

KILOGRAM-METERS TO FOOT-POUNDS.

Kilogram-Meters.	Foot-Pounds.	Kilogram-Meters.	Foot-Pounds.
1	7 233	26	188.058
2	14.466	27	195.291
3	21.699	28	202.524
4	28.932	29	209.757
5	36 165	30	216.990
6	43.398	31	224.223
7	50.631	32	231.456
8	57.864	33	238.689
9	65.097	34	245.922
10	72.330	35	253.155
11	79.563	36	260 388
12	86 796	37	267.621
13	94.029	38	274.854
14	101.262	39	282.087
15	108.495	40	289.320
16	115.728	41	296.553
17	122.961	42	303.786
18	130.194	43	311.019
19	137.427	44	318.252
20	144.660	45	325.485
21	151.893	46	332.718
22	159.126	47	339.951
23	166.359	48	347.184
24	173.592	49	354.417
25	180.825	50	361.650

KILOGRAM-METERS TO FOOT-POUNDS.

Kilogram-Meters.	Foot-Pounds.	Kilogram-Meters.	Foot-Pounds.
51	368.883	76	549.709
52	376.116	77	556.942
53	383.349	78	564.175
54	390.582	79	571.408 ̗
55	397.815	80	578.641
56	405.048	81	585.874
57	412.281	82	593.107
58	419.514	83	600.340
59	426.747	84	607.573
60	433.980	85	614.806
61	441.213	86	622.039
62	448.446	87	629.272
63	455.679	88	636.505
64	462.912	89	643.738
65	470.145	90	650.971
66	477.378	91	658.204
67	484.611	92	665.437
68	491.844	93	672.670
69	499.077	94	679.903
70	506.310	95	687.136
71	513.543	96	694.369
72	520.777	97	701.602
73	528.010	98	708.835
74	535.243	99	716.068
75	542.476	100	723.301

FOOT-POUNDS TO KILOGRAM-METERS.

Foot-Pounds.	Kilogram-Meters.	Foot-Pounds.	Kilogram-Meters.
1	0.13826	26	3.59463
2	0.27651	27	3.73289
3	0.41477	28	3.87114
4	0.55302	29	4.00940
5	0.69128	30	4.14765
6	0 82953	31	4.28591
7	0 96778	32	4.42416
8	1.10604	33	4.56242
9	1.24429	34	4.70067
10	1.38255	35	4.83893
11	1.52081	36	4.97728
12	1.65906	37	5.11544
13	1.79732	38	5.25369
14	1 93557	39	5.39195
15	2 07383	40	5.53020
16	2.21208	41	5.66846
17	2 35034	42	5 80671
18	2.48859	43	5 94497
19	2.62685	44	6 08322
20	2.76510	45	6.22148
21	2.90336	46	6.35973
22	3 04161	47	6.49799
23	3.17987	48	6.63624
24	3.31812	49	6.77450
25	3.45638	50	6.91276

FOOT-POUNDS TO KILOGRAM-METERS.

Foot-Pounds.	Kilogram-Meters.	Foot-Pounds.	Kilogram-Meters.
51	7.05101	76	10.50739
52	7.18927	77	10.64564
53	7.32752	78	10.78390
54	7.46578	79	10.92215
55	7.60403	80	11.06041
56	7.74229	81	11.19866
57	7.88054	82	11.33692
58	8.01880	83	11.47517
59	8.15705	84	11.61343
60	8.29531	85	11.75168
61	8.43356	86	11.88994
62	8.57182	87	12.02819
63	8.71007	88	12.16645
64	8.84833	89	12.30470
65	8.98658	90	12.44296
66	9.12484	91	12.58121
67	9.26309	92	12.71947
68	9.40135	93	12.85772
69	9.53960	94	12.99598
70	9.67786	95	13.13423
71	9.81611	96	13.27249
72	9.95437	97	13.41074
73	10.09262	98	13.54900
74	10.23088	99	13.68725
75	10.36913	100	13.82551

CALORIES TO BRITISH HEAT UNITS.

Calories.	British Heat Units.	Calories.	British Heat Units.
1	3.97	26	103.18
2	7.94	27	107.14
3	11.90	28	111.11
4	15.87	29	115.08
5	19.84	30	119.05
6	23.81	31	123.02
7	27.78	32	126.99
8	31.75	33	130.95
9	35.71	34	134.92
10	39.68	35	138.89
11	43.65	36	142.86
12	47.62	37	146.83
13	51.59	38	150.80
14	55.56	39	154.76
15	59.52	40	158.73
16	63.49	41	162.70
17	67.46	42	166.67
18	71.43	43	170.64
19	75.40	44	174.61
20	79.37	45	178.57
21	83.33	46	182.54
22	87.30	47	186.51
23	91.27	48	190.48
24	95 24	49	194.45
25	99 21	50	198.42

CALORIES TO BRITISH HEAT UNITS.

Calories.	British Heat Units.	Calories.	British Heat Units.
51	202.38	76	301.59
52	206.35	77	305.56
53	210.32	78	309.53
54	214.29	79	313.50
55	218.26	80	317.47
56	222.23	81	321.43
57	226.19	82	325.40
58	230.16	83	329.37
59	234.13	84	333.34
60	238.10	85	337.31
61	242.07	86	341.28
62	246.04	87	345.24
63	250.00	88	349.21
64	253.97	89	353.18
65	257.94	90	357.15
66	261.91	91	361.12
67	265.88	92	365.09
68	269.85	93	369.05
69	273.81	94	373.02
70	277.78	95	376.99
71	281.75	96	380.96
72	285.72	97	384.93
73	289.69	98	388.90
74	293.66	99	392.86
75	297.62	100	396.83

BRITISH HEAT UNITS TO CALORIES.

British Heat Units.	Calories.	British Heat Units.	Calories.
1	0.252	26	6.552
2	0.504	27	6.804
3	0.756	28	7.056
4	1.008	29	7.308
5	1.260	30	7.560
6	1.512	31	7.812
7	1.764	32	8.064
8	2.016	33	8.316
9	2.268	34	8.568
10	2.520	35	8.820
11	2.772	36	9.072
12	3.024	37	9.324
13	3.276	38	9.576
14	3.528	39	9.828
15	3.780	40	10.080
16	4.032	41	10.332
17	4.284	42	10.584
18	4.536	43	10.836
19	4.788	44	11.088
20	5.040	45	11.340
21	5.292	46	11.592
22	5.544	47	11.844
23	5.796	48	12.096
24	6.048	49	12.348
25	6.300	50	12.600

BRITISH HEAT UNITS TO CALORIES.

British Heat Units.	Calories.	British Heat Units.	Calories.
51	12.852	76	19.152
52	13.104	77	19.404
53	13.356	78	19.656
54	13.608	79	19 908
55	13.860	80	20.160
56	14.112	81	20.412
57	14.364	82	20.664
58	14.616	83	20.916
59	14.868	84	21.168
60	15.120	85	21.420
61	15.372	86	21.672
62	15.624	87	21.924
63	15.876	88	22.176
64	16.128	89	22.428
65	16.380	90	22.680
66	16.632	91	22.932
67	16.884	92	23.184
68	17.136	93	23.436
69	17.388	94	23.688
70	17.640	95	23.940
71	17.892	96	24.192
72	18.144	97	24.444
73	18.396	98	24.696
74	18.648	99	24.948
75	18.900	100	25.200

CALORIES TO FOOT POUNDS.

Calories.	Foot Pounds.	Calories.	Foot Pounds.
1	3,091.	26	80,375.
2	6,183.	27	83,467.
3	9,274.	28	86,558.
4	12,365.	29	89,649.
5	15,457.	30	92,741.
6	18,548.	31	95,832.
7	21,640.	32	98,924.
8	24,731.	33	102,015.
9	27,822.	34	105,106.
10	30,914.	35	108,198.
11	34,005.	36	111,289.
12	37,096.	37	114,380.
13	40,188.	38	117,472.
14	43,279.	39	120,563.
15	46,370.	40	123,654.
16	49,462.	41	126,746.
17	52,553.	42	129,837.
18	55,644.	43	132,928.
19	58,736.	44	136,020.
20	61,827.	45	139,111.
21	64,919.	46	142,203.
22	68,010.	47	145,294.
23	71,101.	48	148,385.
24	74,193.	49	151,477.
25	77,284.	50	154,568.

CALORIES TO FOOT-POUNDS.

Calories.	Foot-Pounds.	Calories.	Foot-Pounds.
51	157,659.	76	234,943.
52	160,751.	77	238,035.
53	163,842.	78	241,126.
54	166,933.	79	244,217.
55	170,025.	80	247,309.
56	173,116.	81	250,400.
57	176,208.	82	253,492.
58	179,299.	83	256,583.
59	182,390.	84	259,674.
60	185,482.	85	262,766.
61	188,573.	86	265,857.
62	191,664.	87	268,948.
63	194,756.	88	272,040.
64	197,847.	89	275,131.
65	200,938.	90	278,222.
66	204,030.	91	281,314.
67	207,121.	92	284,405.
68	210,212.	93	287,496.
69	213,304.	94	290,588.
70	216,395.	95	293,679.
71	219,487.	96	296,771.
72	222 578.	97	299,862.
73	225,669.	98	302,953.
74	228,761.	99	306,045.
75	231,852.	100	309,136.

FOOT-POUNDS TO CALORIES.

Foot-Pounds.	Calories.	Foot-Pounds.	Calories.
1	.000 323	26	.008 410
2	.000 647	27	.008 734
3	.000 970	28	.009 057
4	.001 294	29	.009 381
5	.001 617	30	.009 704
6	.001 941	31	.010 028
7	.002 264	32	.010 351
8	.002 588	33	.010 675
9	.002 911	34	.010 998
10	.003 235	35	.011 322
11	.003 558	36	.011 645
12	.003 882	37	.011 969
13	.004 205	38	.012 292
14	.004 529	39	.012 616
15	.004 852	40	.012 939
16	.005 176	41	.013 263
17	.005 499	42	.013 586
18	.005 823	43	.013 910
19	.006 146	44	.014 233
20	.006 470	45	.014 557
21	.006 793	46	.014 880
22	.007 117	47	.015 204
23	.007 440	48	.015 527
24	.007 764	49	.015 851
25	.008 087	50	.016 174

FOOT-POUNDS TO CALORIES.

Foot-Pounds.	Calories.	Foot-Pounds.	Calories.
51	.016 497	76	.024 584
52	.016 821	77	.024 908
53	.017 144	78	.025 231
54	.017 468	79	.025 555
55	.017 791	80	.025 878
56	.018 115	81	.026 202
57	.018 438	82	.026 525
58	.018 762	83	.026 849
59	.019 085	84	.027 172
60	.019 409	85	.027 496
61	.019 732	86	.027 819
62	.020 056	87	.028 143
63	.020 379	88	.028 466
64	.020 703	89	.028 790
65	.021 026	90	.029 113
66	.021 350	91	.029 437
67	.021 673	92	.029 760
68	.021 997	93	.030 084
69	.022 320	94	.030 407
70	.022 644	95	.030 731
71	.022 967	96	.031 054
72	.023 291	97	.031 378
73	.023 614	98	.031 701
74	.023 938	99	.032 025
75	.024 261	100	.032 348

FORCES DE CHEVAUX TO HORSE-POWER.

Forces de Chevaux.	Horse-power.	Forces de Chevaux.	Horse-power.
1	0.986	26	25.644
2	1.973	27	26.631
3	2.959	28	27.617
4	3.945	29	28.603
5	4.932	30	29.590
6	5.918	31	30.576
7	6.904	32	31.562
8	7.891	33	32.549
9	8.877	34	33.535
10	9.863	35	34.521
11	10.850	36	35.508
12	11.836	37	36.494
13	12.822	38	37.480
14	13.808	39	38.466
15	14.795	40	39.453
16	15.781	41	40.439
17	16.767	42	41.425
18	17.754	43	42.412
19	18.740	44	43.398
20	19.726	45	44.384
21	20.713	46	45.371
22	21.699	47	46.357
23	22.685	48	47.343
24	23.672	49	48.330
25	24.658	50	49.316

FORCES DE CHEVAUX TO HORSE-POWER.

Forces de Chevaux.	Horse-power.	Forces de Chevaux.	Horse-power.
51	50.302	76	74.960
52	51.289	77	75.947
53	52.275	78	76.933
54	53.261	79	77.919
55	54.248	80	78.906
56	55.234	81	79.892
57	56.220	82	80.878
58	57.207	83	81.865
59	58.193	84	82.851
60	59.179	85	83.837
61	60.166	86	84.824
62	61.152	87	85.810
63	62.138	88	86.796
64	63.124	89	87.782
65	64.111	90	88.769
66	65.097	91	89.755
67	66.083	92	90.741
68	67.070	93	91.728
69	68.056	94	92.714
70	69.042	95	93.700
71	70.029	96	94.687
72	71.015	97	95 673
73	72.001	98	96.659
74	72.988	99	97.646
75	73.974	100	98.632

HORSE-POWER TO FORCES DE CHEVAUX.

Horse-power.	Forces de Chevaux.	Horse-power.	Forces de Chevaux.
1	1.014	26	26.361
2	2.028	27	27.374
3	3.042	28	28.388
4	4.055	29	29 402
5	5.069	30	30.416
6	6.083	31	31.430
7	7.097	32	32.444
8	8.111	33	33.458
9	9.125	34	34.472
10	10.139	35	35.485
11	11.153	36	36.499
12	12.166	37	37.513
13	13.180	38	38.527
14	14.194	39	39 541
15	15.208	40	40.555
16	16.222	41	41.569
17	17.236	42	42.583
18	18.250	43	43.596
19	19.264	44	44.610
20	20.277	45	45.624
21	21.291	46	46.638
22	22.305	47	47.652
23	23.319	48	48.666
24	24 333	49	49.680
25	25.347	50	50.694

HORSE-POWER TO FORCES DE CHEVAUX.

Horse-Power.	Forces de Chevaux.	Horse-Power.	Forces de Chevaux.
51	51.707	76	77.054
52	52.721	77	78.068
53	53.735	78	79.082
54	54.749	79	80.096
55	55.763	80	81.110
56	56.777	81	82.123
57	57.791	82	83.137
58	58.804	83	84.151
59	59.818	84	85.165
60	60.832	85	86.179
61	61.846	86	87.193
62	62.860	87	88.207
63	63.874	88	89.221
64	64.888	89	90.234
65	65.902	90	91.248
66	66.915	91	92.262
67	67.929	92	93.276
68	68.943	93	94.290
69	69.957	94	95.304
70	70.971	95	96.318
71	71.985	96	97.332
72	72.999	97	98.345
73	74.013	98	99.359
74	75.026	99	100.373
75	76.040	100	101.387

GRAMMES IN A CUBIC CENTIMETER TO OUNCES IN A CUBIC INCH.

Grammes in a Cubic Centimeter.	Ounces in a Cubic Inch.	Grammes in a Cubic Centimeter.	Ounces in a Cubic Inch.
1	0.578	26	15.029
2	1.156	27	15.607
3	1.734	28	16.185
4	2.312	29	16.763
5	2.890	30	17.341
6	3 468	31	17.919
7	4.046	32	18.497
8	4.624	33	19.075
9	5.202	34	19 653
10	5.780	35	20.231
11	6.358	36	20.809
12	6.936	37	21.387
13	7.515	38	21.966
14	8.093	39	22.544
15	8.671	40	23.122
16	9.249	41	23.700
17	9.827	42	24.278
18	10.405	43	24.856
19	10.983	44	25.434
20	11.561	45	26.012
21	12.139	46	26.590
22	12.717	47	27.168
23	13.295	48	27 746
24	13.873	49	28.324
25	14.451	50	28.902

GRAMMES IN A CUBIC CENTIMETER TO OUNCES IN A CUBIC INCH.

Grammes in a Cubic Centimeter.	Ounces in a Cubic Inch.	Grammes in a Cubic Centimeter.	Ounces in a Cubic Inch.
51	29.480	76	43.931
52	30.058	77	44.509
53	30.636	78	45.087
54	31.214	79	45.665
55	31.792	80	46.243
56	32 370	81	46.821
57	32.948	82	47.399
58	33.526	83	47.977
59	34.104	84	48.555
60	34.682	85	49.133
61	35.260	86	49.711
62	35.838	87	50.289
63	36.417	88	50.868
64	36.995	89	51.446
65	37.573	90	52.024
66	38.151	91	52.602
67	38.729	92	53.180
68	39.307	93	53.758
69	39.885	94	54.336
70	40.463	95	54.914
71	41.041	96	55.492
72	41.619	97	56.070
73	42.197	98	56 648
74	42.775	99	57.226
75	43.353	100	57.804

OUNCES IN A CUBIC INCH TO GRAMMES IN A CUBIC CENTIMETER.

Ounces in a Cubic Inch.	Grammes in a Cubic Centimeter.	Ounces in a Cubic Inch.	Grammes in a Cubic Centimeter.
1	1.730	26	44.980
2	3.460	27	46.710
3	5.190	28	48.440
4	6.920	29	50.170
5	8.650	30	51.900
6	10.380	31	53.630
7	12.110	32	55.360
8	13.840	33	57.090
9	15.570	34	58.820
10	17.300	35	60.550
11	19.030	36	62.280
12	20.760	37	64.010
13	22.490	38	65.740
14	24.220	39	67.470
15	25.950	40	69.200
16	27.680	41	70.930
17	29.410	42	72.660
18	31.140	43	74.390
19	32.870	44	76.120
20	34.600	45	77.850
21	36.330	46	79.580
22	38.060	47	81.310
23	39.790	48	83.040
24	41.520	49	84.770
25	43.250	50	86.500

OUNCES IN A CUBIC INCH TO GRAMMES IN A CUBIC CENTIMETER.

Ounces in a Cubic Inch.	Grammes in a Cubic Centimeter.	Ounces in a Cubic Inch.	Grammes in a Cubic Centimeter.
51	88.229	76	131.479
52	89 959	77	133.209
53	91.689	78	134.939
54	93.419	79	136.669
55	95.149	80	138.399
56	96.879	81	140.129
57	98.609	82	141.859
58	100.339	83	143.589
59	102.069	84	145.319
60	103.799	85	147.049
61	105.529	86	148.779
62	107.259	87	150.509
63	108.989	88	152.239
64	110.719	89	153.969
65	112.449	90	155.699
66	114.179	91	157.429
67	115.909	92	159.159
68	117.639	93	160.889
69	119.369	94	162.619
70	121.099	95	164.349
71	122.829	96	166.079
72	124.559	97	167.809
73	126.289	98	169.539
74	128.019	99	171.269
75	129.749	100	172.999

KILOGRAMS IN A CUBIC METER TO POUNDS IN A CUBIC FOOT.

Kilograms in a Cubic Meter.	Pounds in a Cubic Foot.	Kilograms in a Cubic Meter.	Pounds in a Cubic Foot.
1	0.0624	26	1.6231
2	0.1249	27	1.6856
3	0.1873	28	1.7480
4	0.2497	29	1.8104
5	0.3121	30	1.8728
6	0.3746	31	1.9353
7	0.4370	32	1.9977
8	0.4994	33	2.0601
9	0.5619	34	2.1226
10	0.6243	35	2.1850
11	0.6867	36	2.2474
12	0.7491	37	2.3098
13	0.8116	38	2.3723
14	0.8740	39	2.4347
15	0.9364	40	2.4971
16	0.9988	41	2.5595
17	1.0613	42	2.6220
18	1.1237	43	2.6844
19	1.1861	44	2.7468
20	1.2486	45	2.8093
21	1.3110	46	2.8717
22	1.3734	47	2.9341
23	1.4358	48	2.9965
24	1.4983	49	3.0590
25	1.5607	50	3.1214

KILOGRAMS IN A CUBIC METER TO POUNDS IN A CUBIC FOOT.

Kilograms in a Cubic Meter.	Pounds in a Cubic Foot.	Kilograms in a Cubic Meter.	Pounds in a Cubic Foot.
51	3.1838	76	4.7445
52	3.2463	77	4.8070
53	3.3087	78	4.8694
54	3.3711	79	4.9318
55	3.4335	80	4.9942
56	3.4960	81	5.0567
57	3.5584	82	5.1191
58	3.6208	83	5.1815
59	3.6833	84	5.2440
60	3.7457	85	5.3064
61	3.8081	86	5.3688
62	3.8705	87	5.4312
63	3.9330	88	5.4937
64	3.9954	89	5.5561
65	4.0578	90	5.6185
66	4.1202	91	5.6809
67	4.1827	92	5.7434
68	4.2451	93	5.8058
69	4.3075	94	5.8682
70	4.3700	95	5.9307
71	4.4324	96	5.9931
72	4.4948	97	6.0555
73	4.5572	98	6.1179
74	4.6197	99	6.1804
75	4.6821	100	6.2428

POUNDS IN A CUBIC FOOT TO KILOGRAMS IN A CUBIC METER.

Pounds in a Cubic Foot.	Kilograms in a Cubic Meter.	Pounds in a Cubic Foot.	Kilograms in a Cubic Meter.
1	16.02	26	416.48
2	32.04	27	432.50
3	48.06	28	448.52
4	64.07	29	464.54
5	80.09	30	480.56
6	96.11	31	496.57
7	112.13	32	512.59
8	128.15	33	528.61
9	144.17	34	544.63
10	160.18	35	560.65
11	176.20	36	576.67
12	192.22	37	592.68
13	208.24	38	608.70
14	224.26	39	624.72
15	240.28	40	640.74
16	256.30	41	656.76
17	272.31	42	672.78
18	288.33	43	688.80
19	304.35	44	704.81
20	320.37	45	720.83
21	336.39	46	736.85
22	352.41	47	752.87
23	368.43	48	768.89
24	384.44	49	784.91
25	400.46	50	800.92

POUNDS IN A CUBIC FOOT TO KILOGRAMS IN A CUBIC METER.

Pounds in a Cubic Foot.	Kilograms in a Cubic Meter.	Pounds in a Cubic Foot.	Kilograms in a Cubic Meter.
51	816.94	76	1217.41
52	832.96	77	1233.42
53	848.98	78	1249.44
54	865.00	79	1265.46
55	881.02	80	1281.48
56	897.04	81	1297.50
57	913.05	82	1313.52
58	929.07	83	1329.54
59	945.09	84	1345.55
60	961.11	85	1361.57
61	977.13	86	1377.59
62	993.15	87	1393.61
63	1009.17	88	1409.63
64	1025.18	89	1425.65
65	1041.20	90	1441.66
66	1057.22	91	1457.68
67	1073.24	92	1473.70
68	1089.26	93	1489.72
69	1105.28	94	1505.74
70	1121.30	95	1521.76
71	1137.31	96	1537.78
72	1153.33	97	1553.79
73	1169.35	98	1569.81
74	1185.37	99	1585.83
75	1201.39	100	1601.85

MILLIERS IN A CUBIC METER TO SHORT TONS IN A CUBIC YARD.

Milliers in a Cubic Meter.	Short Tons in a Cubic Yard.	Milliers in a Cubic Meter.	Short Tons in a Cubic Yard
1	0.843	26	21.912
2	1.686	27	22.755
3	2.528	28	23.598
4	3.371	29	24.441
5	4 214	30	25 283
6	5.057	31	26.126
7	5.899	32	26.969
8	6.742	33	27.812
9	7.585	34	28.654
10	8.428	35	29.497
11	9.271	36	30 340
12	10.113	37	31.183
13	10.956	38	32.026
14	11.799	39	32.868
15	12.642	40	33.711
16	13.484	41	34.554
17	14.327	42	35.397
18	15.170	43	36.239
19	16.013	44	37.082
20	16.856	45	37.925
21	17.698	46	38.768
22	18.541	47	39.611
23	19.384	48	40.453
24	20.227	49	41.296
25	21.069	50	42.139

MILLIERS IN A CUBIC METER TO SHORT TONS IN A CUBIC YARD.

Milliers in a Cubic Meter.	Short Tons in a Cubic Yard.	Milliers in a Cubic Meter.	Short Tons in a Cubic Yard.
51	42.982	76	64.051
52	43.824	77	64.894
53	44.667	78	65.737
54	45.510	79	66.579
55	46 353	80	67.422
56	47.196	81	68.265
57	48.038	82	69.108
58	48.881	83	69.951
59	49.724	84	70.793
60	50.567	85	71.636
61	51 409	86	72.479
62	52.252	87	73 322
63	53.095	88	74.164
64	53.938	89	75.007
65	54.781	90	75.850
66	55.623	91	76.693
67	56.466	92	77.536
68	57 309	93	78.378
69	58.152	94	79.221
70	58.994	95	80.064
71	59.837	96	80.907
72	60.680	97	81.749
73	61.523	98	82.592
74	62.366	99	83.435
75	63 208	100	84.278

SHORT TONS IN A CUBIC YARD TO MILLIERS IN A CUBIC METER.

Short Tons in a Cubic Yard.	Milliers in a Cubic Meter.	Short Tons in a Cubic Yard	Milliers in a Cubic Meter.
1	1.187	26	30.850
2	2.373	27	32.037
3	3.560	28	33.223
4	4.746	29	34.410
5	5.933	30	35 597
6	7.119	31	36.783
7	8.306	32	37.970
8	9.492	33	39.156
9	10.679	34	40.343
10	11.866	35	41.529
11	13 052	36	42 716
12	14.239	37	43.902
13	15.425	38	45.089
14	16.612	39	46.275
15	17.798	40	47.462
16	18.985	41	48 649
17	20.171	42	49.835
18	21.358	43	51.022
19	22.544	44	52.208
20	23.731	45	53.395
21	24.918	46	54.581
22	26.104	47	55.768
23	27.291	48	56.954
24	28.477	49	58.141
25	29.664	50	59.328

SHORT TONS IN A CUBIC YARD TO MILLIERS IN A CUBIC METER.

Short Tons in a Cubic Yard.	Milliers in a Cubic Meter.	Short Tons in a Cubic Yard.	Milliers in a Cubic Meter.
51	60.514	76	90 178
52	61.701	77	91.364
53	62.887	78	92.551
54	64.074	79	93.737
55	65.260	80	94 924
56	66.447	81	96.111
57	67.633	82	97.297
58	68.820	83	98.484
59	70.006	84	99.670
60	71.193	85	100.857
61	72.380	86	102.043
62	73.566	87	103.230
63	74.753	88	104.416
64	75 939	89	105.603
65	77.126	90	106.790
66	78.312	91	107.976
67	79.499	92	109.163
68	80.685	93	110.349
69	81.872	94	111.536
70	83.058	95	112.722
71	84.245	96	113.909
72	85.432	97	115.095
73	86.618	98	116 282
74	87.805	99	117.468
75	88.991	100	118.655

MILLIGRAMS IN A LITER TO GRAINS IN A U. S. GALLON.

Milligrams in a Liter.	Grains in a U. S. Gallon.	Milligrams in a Liter.	Grains in a U. S. Gallon.
1	0.058	26	1.519
2	0.117	27	1.578
3	0.175	28	1.636
4	0.234	29	1.695
5	0.292	30	1.753
6	0.351	31	1.812
7	0.409	32	1.870
8	0.468	33	1.929
9	0.526	34	1.987
10	0.584	35	2.045
11	0.643	36	2.104
12	0.701	37	2.162
13	0.760	38	2.221
14	0.818	39	2.279
15	0.877	40	2.338
16	0.935	41	2.396
17	0.993	42	2.454
18	1.052	43	2.513
19	1.110	44	2.571
20	1.169	45	2.630
21	1.227	46	2.688
22	1.286	47	2.747
23	1.344	48	2.805
24	1.403	49	2.864
25	1.461	50	2.922

MILLIGRAMS IN A LITER TO GRAINS IN A U. S. GALLON.

Milligrams in a Liter.	Grains in a U. S. Gallon.	Milligrams in a Liter.	Grains in a U. S. Gallon.
51	2.980	76	4.441
52	3.039	77	4.500
53	3.097	78	4.558
54	3.156	79	4.617
55	3.214	80	4 675
56	3.273	81	4.734
57	3.331	82	4.792
58	3.390	83	4.851
59	3.448	84	4 909
60	3.506	85	4.967
61	3.565	86	5.026
62	3.623	87	5.084
63	3.682	88	5.143
64	3.740	89	5.201
65	3.799	90	5.260
66	3.857	91	5.318
67	3.915	92	5.376
68	3.974	93	5.435
69	4.032	94	5.493
70	4.091	95	5.552
71	4.149	96	5.610
72	4.208	97	5.669
73	4.266	98	5.727
74	4.325	99	5.786
75	4.383	100	5.844

GRAINS IN A U. S. GALLON TO MILLIGRAMS IN A LITER.

Grains in a U. S. Gallon.	Milligrams in a Liter.	Grains in a U. S. Gallon.	Milligrams in a Liter.
1 .	17.1	26	444.9
2	·34.2	27	462.0
3	51.3	28	479.1
4	68.4	29	496.2
5	85.6	30	513.4
6	102.7	31	530.5
7	119.8	32	547.6
8	136.9	33	564.7
9	154.0	34	581.8
10	171.1	35	598.9
11	188.2	36	616.0
12	205.3	37	633.1
13	222.5	38	650.3
14	239.6	39	667.4
15	256.7	40	684.5
16	273.8	41	701.6
17	290.9	42	718.7
18	308.0	43	735.8
19	325.1	44	752.9
20	342.2	45	770.0
21	359.4	46	787.2
22	376.5	47	804.3
23	393 6	48	821.4
24	410.7	49	838.5
25	427.8	50	855.6

GRAINS IN A U. S. GALLON TO MILLIGRAMS IN A LITER.

Grains in a U. S. Gallon.	Milligrams in a Liter.	Grains in a U. S. Gallon.	Milligrams in a Liter.
51	872.7	76	1300.5
52	889.8	77	1317.6
53	906.9	78	1334.7
54	924.0	79	1351.8
55	941.2	80	1369.0
56	958.3	81	1386.1
57	975.4	82	1403.2
58	992.5	83	1420.3
59	1009.6	84	1437.4
60	1026.7	85	1454.5
61	1043.8	86	1471.6
62	1060.9	87	1488.7
63	1078.1	88	1505.9
64	1095 2	89	1523.0
65	1112.3	90	1540.1
66	1129.4	91	1557.2
67	1146.5	92	1574.3
68	1163.6	93	1591.4
69	1180.7	94	1608.5
70	1197.8	95	1625.6
71	1215.0	96	1642.8
72	1232.1	97	1659.9
73	1249.2	98	1677.0
74	1266.3	99	1694.1
75	1283 4	100	1711.2

COMPARISON OF THERMOMETER SCALES.

Reading of Centigrade Thermometer.	Reading of Fahrenheit Thermometer.	Reading of Centigrade Thermometer.	Reading of Fahrenheit Thermometer.	Reading of Centigrade Thermometer.	Reading of Fahrenheit Thermometer.
—20°	—4°	105°	221°	230°	446°
—15	+5	110	230	240	464
—10	14	115	239	250	482
—5	23	120	248	260	500
0	32	125	257	270	518
+5	41	130	266	280	536
10	50	135	275	290	554
15	59	140	284	300	572
20	68	145	293	310	590
25	77	150	302	320	608
30	86	155	311	330	626
35	95	160	320	340	644
40	104	165	329	350	662
45	113	170	338	360	680
50	122	175	347	370	698
55	131	180	356	380	716
60	140	185	365	390	734
65	149	190	374	400	752
70	158	195	383	410	770
75	167	200	392	420	788
80	176	205	401	430	806
85	185	210	410	440	824
90	194	215	419	450	842
95	203	220	428	460	860
100	212	225	437	470	878

AUXILIARY TABLE.

Centigrade.	Fahrenheit.	Fahrenheit.	Centigrade.
0°	0.0°	0°	0.00°
1	1.8	1	0.56
2	3.6	2	1.11
3	5.4	3	1.67
4	7.2	4	2.22
5	9.0	5	2.78
6	10.8	6	3.33
7	12.6	7	3.89
8	14.4	8	4.44
9	16.2	9	5.00
10	18.0	10	5.56
11	19.8	11	6.11
12	21.6	12	6.67
13	23.4	13	7.22
14	25.2	14	7.78
15	27.0	15	8.33
16	28.8	16	8.89
17	30.6	17	9.44
18	32.4	18	10.00
19	34.2	19	10.56
20	36.0	20	11.11

CAUTION.—In comparing readings on the Centigrade and Fahrenheit thermometer scales, use the table on the *opposite* page. The table on *this* page is only a supplementary one. Its use is explained on page 35.

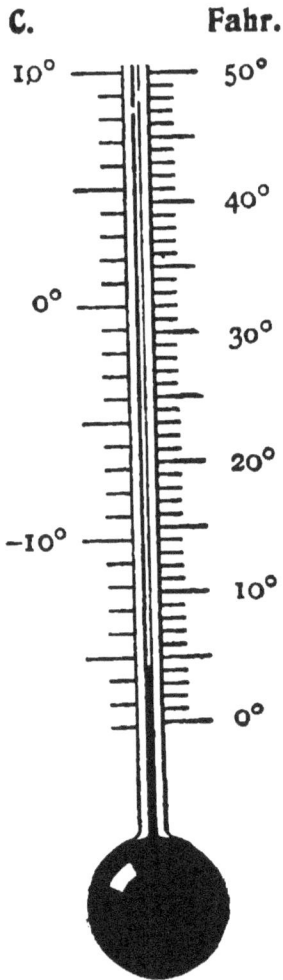

Comparison
of
THERMOMETER
SCALES.

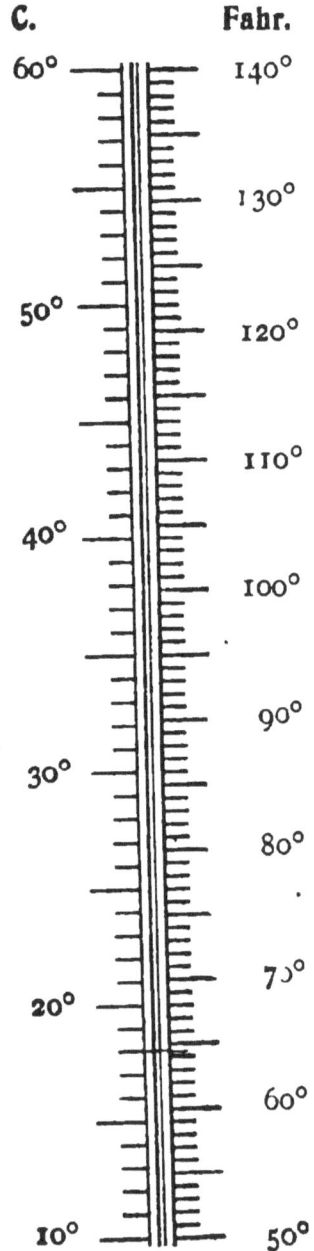

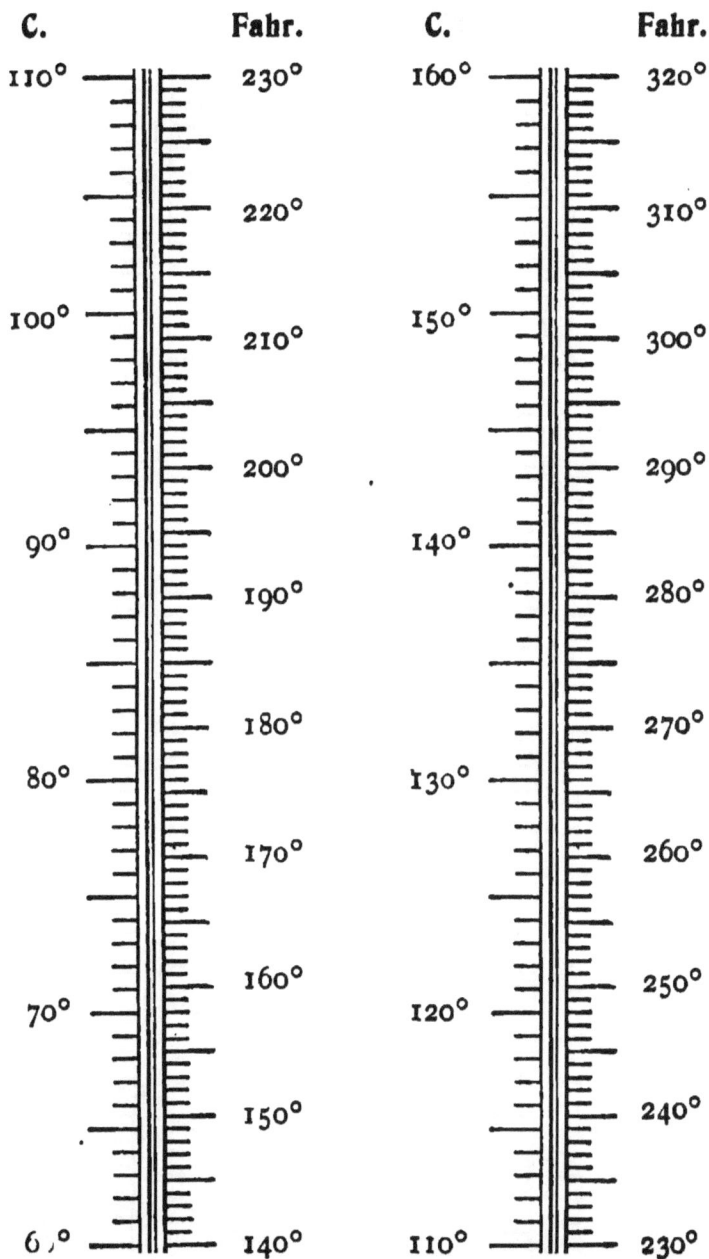

C.	Fahr.	C.	Fahr.
110°	230°	160°	320°
	220°		310°
100°	210°	150°	300°
	200°		290°
90°	190°	140°	280°
	180°		270°
80°	170°	130°	260°
	160°	120°	250°
70°	150°		240°
60°	140°	110°	230°

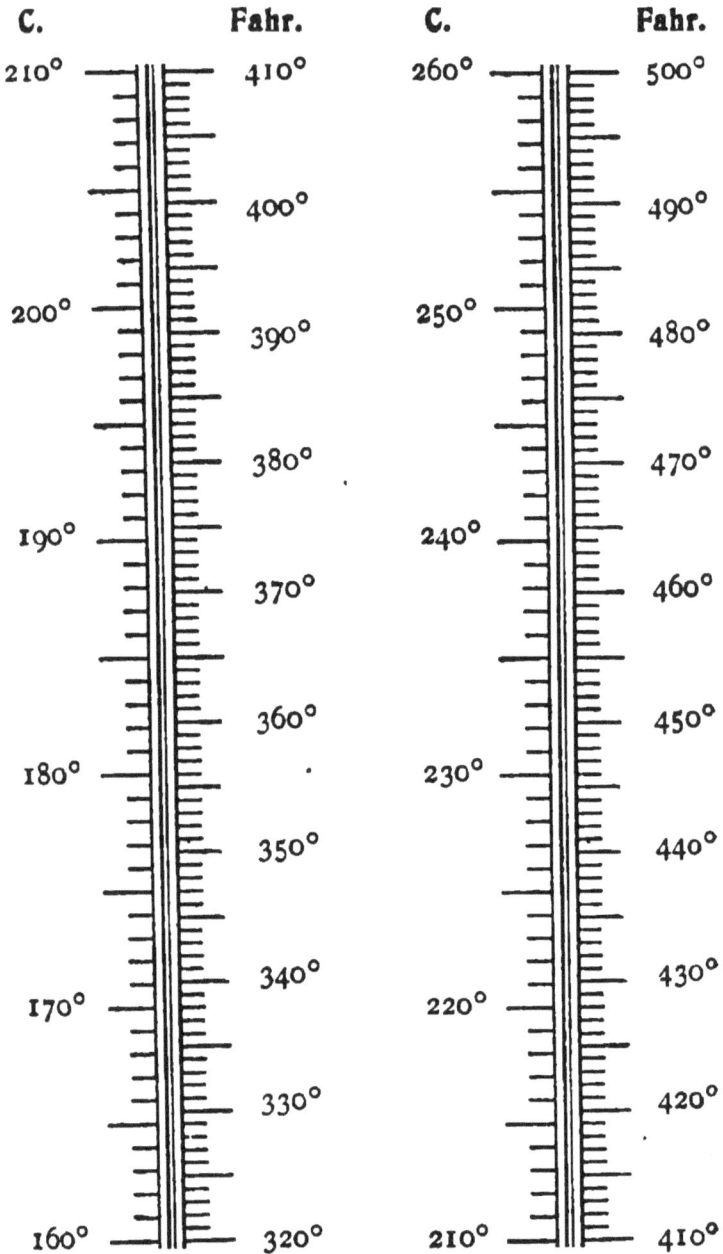

C.	Fahr.	C.	Fahr.
210°	410°	260°	500°
	400°		490°
200°	390°	250°	480°
	380°		470°
190°	370°	240°	460°
	360°		450°
180°	350°	230°	440°
	340°		430°
170°	330°	220°	420°
160°	320°	210°	410°

INDEX TO TABLES.

Acres to hectares, 78.
Avoirdupois ounces to grammes, 130.
            pounds to kilograms, 134.
British gallons to liters, 114.
      heat units to calories, 162.
      liquid quarts to liters, 110.
Bushels, U. S., to steres, 122.
Calories to British heat units, 160.
      to foot-pounds, 164.
Centigrade.—See *Thermometer*.
Centimeters to inches, 44.
Chevaux, forces de, to horse power, 168.
Cubic centimeters to cubic inches, 80.
         to U. S. fluid ounces, 96.
   feet to cubic meters, 86.
   inches to cubic centimeters, 82.
   kilometers to cubic miles, 92.
   meters to cubic feet, 84.
      to cubic yards, 88.
   miles to cubic kilometers, 94.
   yards to cubic meters, 90.
Dry.—See *Liters*, *Quarts*, *Bushels*, and *Steres*.
English.—See *British*.
Fahrenheit.—See *Thermometer*.
Feet to meters, 50.
Fluid ounces to cubic centimeters, 98.

(193)

Foot-pounds to calories, 166.
   to kilogram-meters, 158.
Forces de chevaux to horse power, 168.
Gallons, British, to liters, 114.
   U. S., to liters, 106.
Grains to grammes, 126.
   in a U. S. gallon to milligrams in a liter, 186.
Grammes to grains, 124.
   to avoirdupois ounces, 128.
   to Troy ounces, 136.
   in a cubic centimeter to ounces in a cubic
    inch, 172.
Heat units, British, to calories, 162.
Hectares to acres, 76.
Horse power to forces de chevaux, 170.
Inches to centimeters, 46.
Kilogram-meters to foot-pounds, 156.
Kilograms to avoirdupois pounds, 132.
   to Troy pounds, 140.
   in a cubic meter to pounds in a cubic
    foot, 176.
   per square centimeter to pounds per
    square inch, 152.
   per square meter to pounds per square
    foot, 148.
Kilometers to miles, 56.
Liquid.—See *Quarts*, *Gallons*, and *Liters*.
Liters to British gallons, 112.
   to British liquid quarts, 108.
   to U. S. liquid quarts, 100.
   to U. S. gallons, 104.
   to U. S. dry quarts, 116.
Meters to feet, 48.
   to yards, 52.
Miles to kilometers, 58.

Milliers to short tons, 144.

    in a cubic meter to tons in a cubic yard, 180.

Milligrams in a liter to grains in a U. S. gallon, 184.

Ounces, avoirdupois, to grammes, 130.

    U. S. fluid, to cubic centimeters, 98.

    Troy, to grammes, 138.

    in a cubic inch to grammes in a cubic centimeter, 174.

Pounds, avoirdupois, to kilograms, 134.

    Troy, to kilograms, 142.

    per square inch to kilograms per square centimeter, 154.

    per square foot to kilograms per square meter, 150.

    in a cubic foot to kilograms in a cubic meter, 178.

Quarts, British liquid, to liters, 110.

    U. S. liquid, to liters, 102.

    U. S. dry, to liters, 118.

Scales, thermometer, comparison of, 188, 189, 190.

Short tons to milliers, 146.

    in a cubic yard to milliers in a cubic meter, 182.

Square centimeters to square inches, 60.

    feet to square meters, 66.

    inches to square centimeters, 62.

    kilometers to square miles, 72.

    meters to square feet, 64.

        to square yards, 68.

    miles to square kilometers, 74.

    yards to square meters, 70.

Steres to U. S. bushels, 120.

Temperature.—See *Thermometer.*

Thermometer scales, comparison of, 188, 189, 190.

Tonnes and Tonneaux.—See *Milliers.*

Tons, short, to milliers, 146.
Tons in a cubic yard to milliers in a cubic meter,
    182.
Troy ounces to grammes, 138.
      pounds to kilograms, 142.
United States bushels to steres, 122.
            dry quarts to liters, 118.
            fluid ounces to cubic centimeters, 98.
            gallons to liters, 106.
            liquid quarts to liters, 102.
Yards to meters, 54.

www.ingramcontent.com/pod-product-compliance
Lightning Source LLC
Chambersburg PA
CBHW021800190326

41518CB00007B/388